They Called, We Came

They Called, We Came

Scott D. Snyder

To order additional copies of this book, contact:
Xlibris Corporation
1-888-795-4274
www.Xlibris.com
Orders@Xlibris.com
53673

Prologue

AUGUST 29TH, 2005 a category 4 Hurricane hit the gulf coast and the world changed. Lives were lost, people became scattered and displaced, and the whole country faced a new challenge.

Under the pressure, mistakes were made by federal, state, and local governments leaving many to fend for themselves. The news media shook the world's confidence to the core with its rumors and false accusations. Politicians were busy pointing fingers while public figures took center stage and grandstanded.

Through confusion and hatred there was one virtue that was unshakeable, the human will to help. This is the story of that will, the stories of those who answered the call and went to an unfamiliar, hostile place and made a difference when all seemed lost. We left the safety of our own lives to go save those who had lost everything. This book is our side of the story of what we saw and did. We went into hell and made life and survival our top priority. We shed tears, blood, physical and mental strain to make sure everyone was rescued.

We did what we had to do when others did nothing. For many of us, this experience will never leave us; it is a part of us. As the victims recover from their loss we will never recover from our gain.

This is our story of New Orleans and what happened. Though some of the names in here have been changed the stories are the same. This book is dedicated to the men and women who made a difference and overcame incredible odds. We came from hundreds some even thousands of miles away to help those who needed it most. We couldn't save the world, but I like to think that we saved New Orleans.

Tears to Heroes

I shed the tears for the ones left behind
The families who never got to say the simple words
A last trivial word as hello or just I love you
They are lost, taken when they meant so much.

I shed the tears for the ones lost forever
Lost in the ripples as all hope is lost
Remembered by name forever etched in our memories
Never to be seen again still never forgotten.

I shed the tears for the ones who were heroes
The uniforms they wore and the oath they had taken
Died for what they believed without a selfish thought
To protect us all from the harm they made the sacrifice.

I shed the tears for the innocents lost
The buses and boats that never came of arrived too late
In the line of life they struggled and fought to survive and never surrender
Never to let defeat to prevent the hands of death from destroying lives.

I shed the tears for all who have suffered
For we know no fate as this is what they deserved
To heaven they go with hands of love and security
I shed my tears to them all, to them from me.

To a loved one who has lost a loved one this is dedicated to their memory may it
live as they did.

It's Not Just a Job

It's not just a job, it's who we are
It's about the lives we touch everyday
Whether from the near end or just a smile
It's never just one, it's about each one of them.

It's not just a job, but what we do
Day in day out, it's never the same thing
Each situation different, each person special
They are important to others, they are a priority to us.

It's not just a job, it's why we are here
To help those sick, dying or just in need of us
They don't need a reason we'll come anyway
It's because of them, we love our life.

It's not just a job, it's the life we choose
Nothing else seemed to work for us
The others didn't give us the joy we wanted
Until the first time we saved a lost soul.

It's not just a job, it's our passion for life
When we see the lives we saved smile back
Shake our hand or give a hug with tears for two
It's not just a job, it's the lives I love
It's a life, a choice, and it's a passion for life.

Call to Duty for Honor

A lot can happen to you in the field
A war can battle through yourself
Balance of the mind and body as one element
This is what makes us the best.

With time to pass as the clock turns
Trying as hard as you can to keep yourself busy
Still realizing in this world you are alone
Shadows playing games with your eyes.

Some of us make phone calls to loved ones back home
While the rare few have no calls to make at all
It's the empty pages that often write the most
Those are the ones who have nothing to lose.

Like the lost souls who just carry on existence
No one to worry about, all ties cut off and left behind
Feeling so bad if their life is last of a lonely funeral
A solemn song of goodbye with a tearful last word.

It's the uniform that says call of duty
Senseless acts they perform day in and day out
Just one moment of silence to those fallen
Keep the pride, faith and spirit alive.

Prisoner of Illusion

A dream for a life that could be so much simpler
Inspiration that had come from the observation
Viewing the purpose of life through another
Seeming to think of a love story gone terribly wrong.

Perception and understanding of two different worlds
Like turning out the lights and still having clear sight
Only to have it clogged by a fog of desolation
Now suddenly your eyes are open, but can't see anything.

A sweet dream that has suddenly become a nightmare
As rational thought overthrown by a powerful feeling
When you are filled with rage, pain, sorrow and anger
The whole picture can be thrown out of proportion.

There are questions we ask that can never be answered
Only determination by God, fate, destiny, or heaven above
Maybe they are answers of truth too painful to accept
In many ways we need to know and others we are better without.

I can't decide my own fate or destiny
Control of my life is a sheer illusion
All misguided faith or passed judgment on myself
Found guilty and sentenced by reason of insanity.

Condemned by the action beyond my comprehension
It's the true reasoning by the nature of insanity
For so many it's a way of life to pursue
By others we are prisoners by the lack of any understanding.

Will to the Light

Some came from afar
Some came from across the street
Others came across the ocean
We all came together.

There was a goal that needed to be achieved
A mission that was set to be accomplished
Towns in ruins an entire world all its own was gone
It needed to be found, all of it needed rescued.

Young and old, white and black it didn't matter
World outside put as aside for the common good
Sacrifice from convenience to put forth courage
We knew why we were there, who cared if the world didn't.

No pride or glory
Heroes none of us really wanted to be
It was about what the job and people asked of us
Quest for fame didn't even exist it was about public success.

Those of us who were there
Saw a beam of light through a shroud of shadows
It never failed, swayed and was powered by our strength
Strength of those who refused to walk away.

This was a letter sent to me by a fellow search and rescue volunteer who was there before my arrival. We became friends and I asked her what she remembered about her time in New Orleans and Louisiana in the aftermath of Katrina, this is what she sent,

"My experience with helping out in New Orleans will be something I never forget. It was so nice to be able to look around and see EMS and emergency response teams from all over the U.S.A. working together.

I responded from Jennings, LA. which is about 3 hours west of N.O. (New Orleans). I have been raised around hurricanes but have never experienced the aftermath of one like Katrina. My fire department sent different crews to many places that Katrina hit. I went with our medical unit for five days to N.O. along with my husband who is on the department as well. I was overwhelmed at first when I arrived at I-10 and Causeway, seeing thousands of people standing under the overpass. Military helicopters landed and departed with patients quickly as well as dropped off people who were rescued from rooftops. While transporting patients to Baton Rouge I heard many stories, I listened well and comforted them the best I could. With never going through the experiences before I could not say, "I understand" I could only imagine the fear everyone felt.

The following week after I returned I departed with a fellow volunteer in one of our tankers for another 5 days, this time to Chamette, St. Bernard parish. It was hard to look around at al the nice homes that were torn apart and once underwater. I helped with search and rescue one day. While searching through each home I prayed I would not find a victim. I also prayed for each family because I saw with my own eyes that they lost everything they owned. Seeing the spray painted words on roof tops by the holes saying, "S.O.S." or" We are ok" touched my heart. It must have been traumatic for each and every one of them.

It wasn't long after I returned that we were preparing for Hurricane Rita. We had a couple of crews still out in Mississippi helping there that I had to cut it short and return. After Rita passed, our community had emergency response as well. It was an emotional time for me. To be out helping a community then quickly turn around and need and receive help here was very overwhelming. Thankfully our area was not hit near as hard as N.O. with Katrina. I will never forget all the people that came to our aide. Thank you to everyone that I worked with in N.O. and thank you to everyone that came and helped in J.D. parish! These experiences will forever be remembered by me.

Julie LeJuene
Jennings, Louisiana

Tuesday August 30, 1230

THERE WAS WATER everywhere it looked like something out of a movie where Mother Nature put us back into our place like an old fashioned ass kicking. Fires were burning out of control, peoples homes were gone by the flames that seemed to have a life all their own. There were homes, two story dwellings where all you could see was their rooftops, as far as the eye could see there was nothing but water and you knew there was nothing left.

Looking through the lens of a video camera the images exploded into your television screen like a slap to the face, vivid images burned into your mind with the worst possible scenarios and images. Watching from 500 miles away I was easily able to minimize the impact it had on me, but seeing it right in front in real life before you, it still hit hard.

It captivated you seeing the water rush into a city that was filled with people and a culture all its own, a world of jazz and blues music that had its own language and people that welcomed you with a smile and a strong drink. A world of celebration and free spirit was gone by an act of God, or so some would think. Lives were gone, cars submerged and houses were swallowed up and ravaged by an element that always helped to bring and preserve life not bring death.

Although many fled and left for safety there are many that refused to stray from what they felt was everything they had. Whether it was a little or a lot to them it was their entire world, their home where they built their lives as families they couldn't just walk away. To many it seemed selfish and stupid to risk it all for your home, but I respect their thinking and I think I might have done the same thing.

Tuesday August 30, 2145

IT FELT LIKE we had been driving all night, rain just wouldn't let up as we were coming through Ohio and it just seemed to take longer to get to where we needed to be. Stopping for diesel seems to be the only thing that allows us to keep our mind off of the whole idea of what we are getting ourselves into here. Standing around talking about anything that comes to mind, meanwhile deep down and seeing the look on everyone's face it's pretty clear of what they are thinking about. Looking out the passenger window so many thoughts were going through my mind with questions that I really didn't want answers to. What was it going to be like? Was I really going to be ready for this? Thoughts like these make me remember my past and how others had relied upon me before.

Thinking about the whole situation as if it were happening all in the blink of an eye. I went from being at home with my family to suddenly heading to a place that had no words for description. The world had been watching this event of Hurricane Katrina from our homes safe and dry able to change the channel at any time. Now here I am a veteran of disaster and death going to a scene of both and at an astronomical proportion.

My experience and training has always prepared me physically, it's the emotional and mental factors that can cause the most damage. I have the scars and enough memories to last me a lifetime, but still I see myself having to say inside my head, "One more time, one more mission you know you aren't done yet." It's like a battle of my own will, the devil and the angel on my shoulders fighting.

Am I crazy for accepting this mission? Absolutely, only crazy people are the ones who can do this type of work. Am I ready for this? Probably not, but who ever really is, no one walks around thinking that no matter what happens they wouldn't be surprised. Your training, knowledge and experience can only take you so far; it's a matter of dedication, motivation, and compassion to know what to do. Do I consider myself a leader? Yes, but the true question I am asking myself is do the men consider me a leader? I can't honestly answer that, I have always tried to have my performance speak for itself and when they look at me, they will see strength and guidance. These are the components of a leader and I figure that when it comes down to it all, they will trust me and follow my lead because they know I could do it without doubt or fear.

This is a war of survival against Mother Nature and so far we have lost the battle. I guess that is what has always made me the kind of emergency medical technician I am today; I won't quit and won't let my teammates quit either. It is who we are and it is what we do, our boss John and other managers picked us for a reason, we are the best at what we do in their eyes and the company knows that we won't give up. We don't know how to give up, we are the elite: the most courageous or the craziest sons of bitches alive!

Wednesday, 8/31 0430

TRIED SLEEPING IN the ambulance but can't do it. Too many thoughts keep going through my head, between my loved ones, my team, and what we are going to see it was just too much to think about. I didn't know what my bosses were thinking when they asked me to go, but I feel a lot of pride to be going. Maybe they think that I can do something down there so many others can't. It's so hard to get a read on the other guys, but I know Carl, like myself is scared of what we will find. More sleepless nights are to come, maybe I'll try to doze off again, but I don't think it'll work.

Not having the ability to sleep can have its good and bad points. It helps with stamina and endurance, but it can prevent me from making sound judgment calls that can mean life or death for me and my teammates. I know none of us are tired, but down time in the back of an ambulance with the lights off will help us. I don't know he prepared we truly are, each man is different and each one will react differently to mental and physical stress as well as fatigue and isolation from ourselves and their loved ones.

Kyle, Bruce, and Gary and myself are family men who kissed our wives and kids goodbye and set forth on this journey. It's always the hardest on the kids because they may not understand what is going on. My kids (twins age 4) keep thinking that I am going to play in lava, instead of water despite several attempts to explain otherwise. They don't understand that Dad is going to be surrounded by pain and death.

Jack I think is the lucky one he doesn't have kids to have to say goodbye and look at their faces as he drives away. Jack is the youngest of the group and never had to face anything like this before. Each man has his own reasons for going, each one has an objective and all of them are good men who I would ordinarily trust with my own life, but how some will do when faced with their own worst fears is another story. It is good to have such a diversified group of men that we are to be the representatives for the company. Young and old, veterans and rookies we have it all that way we can feed and build off of each other. We will be a family, a team that relies on each other for everything.

For one of us this mission becomes personal, Gary's father is in Biloxi, Mississippi one of the other heavily effected areas. We know we will at times have to keep a close eye on him sometimes, but it will benefit both him and us. You are no good to your team or yourself if your world has crumbled around you. Situations like that with no matter what the outcome you have to look at the whole picture, means more than just yourself, but the others around you.

To some with years of experience it can come as a sort of demented type of business as usual. For Kyle, this is education and training being played out. He has seen things that he wouldn't wish upon his worst enemies. Life and death he has been there and knows the effects, in some ways I look to him for guidance despite having a lot of the same traits as him. Kyle is the type of man you want around when things go wrong, he can come up with a plan and easily gets others to follow him. My respect for Kyle has no limits, no borders, just a sense of brotherhood you get from shared experience and have proven yourself. There are two things that get you home from a mission, your training and the men you serve with. With Kyle I have no doubt that we will all come home safe.

In every group you always got the young one who is driven and curious. Jack would best describe those two words. We look at him and see the determination in his eyes and hear it in his voice. I think that is why he will be such a great paramedic, with Justin you see the potential and it makes you glad to be by his side. Ready and willing without hesitation to do whatever is needed. A person like that can help keep you going when you just want to give up. This is his first mission for a large scale mass incident and he knows the score. He goes when we go and knows what to do; it's just a matter of him getting that chance.

There is always one that you look at and you just can't figure out what is on their mind. You see a smile on his face, but his eyes are still sharp and focused. Bruce, he is the one you just never know what he is thinking and it can be a little bit exciting and scary at the same time. Bruce brings the resourceful fire fighting background that is useful and very much essential in a disaster like this. He brings knowledge and experience in fields that we either don't have or have never used. He's a good man who cracks a smile and is motivated to want to be here. He also has the traits of a former military man of discipline and a valued sense of teamwork.

The group brings something more of a higher standard to everything we do and it makes us truly exceptional. We will go in as a team and walk out the same way, each man knows who he is and what is expected from not only ourselves, but of each other. I know we will exceed those expectations and take it to a whole new level.

Wednesday 8/31 1100

GOT SOME SLEEP and have been doing a lot of driving, about 30 miles from the Louisiana state line. Traveling down interstate 55 South we could hear on the radio that it is closed at the state line, but I feel confident that we will be able to get through. Driving on the open road we start to see minor damage, but nothing significant other than down trees and busted up old wooden bill boards. There are tree branches on the side of the road and the hillsides are covered with broken tree trunks, but nothing to indicate anything other than a bad thunderstorm. The air is heavy and humid as the radio deejay tells us it's 90 degrees. All I can really think about is thank God I don't have any respitory problems or I would really be screwed.

Driving down the highway we are passing a convoy of utility trucks, there must be 50-60 of them with license plates of at least 5 different states. It's amazing to see how every resource and aspect is being used to fix this catastrophe. We tried to stop for food at a local fast food place; the power bars we have been eating are just too nasty. We pulled into a shopping plaza with a whole bunch of other cars only to find out the whole place doesn't have any power. The stress is already starting to show on the guys, but it's mostly just being tired and not knowing what's going on. We decided to just pull over and have a little vent session to kind of clear the air and get back on track. After a few minutes we loaded back into the ambulance and head back on course to Baton Rouge.

We continuously try to get in touch with a Mr. Irwin and Acadian Ambulance since they are our point of contact, but no one has any information for us. Acadian from what I understand is one of the largest ambulance service in the state of Louisiana and is supposed to be the primary back up for New Orleans EMS. Since a lot of people from Acadian ambulance have lost everything also, it's hard to tell what will happen to them. It's at the point we really begin to wonder how screwed this thing really is. What have we gotten ourselves into and truly what kind of nightmares are we going to be living? I don't want to think about it. I told Bruce that depending what is going on in New Orleans, we may be given the missions of search and rescue or later body recovery, in all we may no be able to save a lot of lives, but maybe we can bring closure to people's lives.

Wednesday, 8/31 1335

GETTING TO BATON Rouge was a challenge in itself with the traffic that seemed to never end. Trying to remain in a tight two-man convoy seemed to be almost impossible and seeing the obstacles we have to deal with I told Bruce we may have to go lights and siren. I look over to my right and see her sitting in her car holding her rosary beads in her hands draped over the steering wheel. I can't help but think about my own faith and belief in God. You can't help but question God as to why He would allow this to happen? How can anyone believe in Him after seeing and going through all this tragedy? I guess in a way it tests man and his ability to overcome something like this. In some ways this can be compared to September 11th and the World Trade Center, Pentagon and Shanksville. There was death surrounding you, engulfed in sorrow, blood and despair and yet somehow we prevailed as a nation, to rebuild our way of life. Maybe God is like a military drill instructor he breaks you down taking away selfishness and greed to create something better and greater than yourself. This isn't about individual glory or achievement for one person it's about all of us and we are going to win.

Communicating with Kyle in the ambulance 9754, I told him we were going to ride the right side of the road going lights and siren so we could keep our deadline with getting to the command post. I told Kyle we would take point and to stay close behind. I directed Bruce to stay close to the right shoulder, keep it below 15 mph, and to ride both the road and the dirt on the side of the road. I figure that way if we came across any hazards we had room to maneuver. The last thing we needed was to blow a tire before we reached the command post.

We came to a shopping plaza right on the main road about five minutes from the command post. It was amazing to find the whole area had no indications that a hurricane even came through there. The shops were open, power was on and people seem to be going on about their daily lives. We decided to stop at a burger joint and gab something to eat. We didn't know when we would get to eat next, so it was best that we eat now before we set out for a long period of time.

While waiting a few police officers were standing in line with us talking about all the violence that was going on in the city. I asked him what was going on and he told that people have been shooting at service vehicles (police, fire, and ambulances) and one police officer had already been shot. Hearing this got me to thinking, "What in the hell have I gotten myself into?" We drove down the main road (Bluebonnet Dr.) and found our turn off titled Ministry Dr. and became instantly engulfed into buildings and emergency vehicles. The complex was some local college and now it was transformed into our Command Center. Kyle and I told the guys to stay with the ambulances while we went inside to check in and get orders. Walking into the main building it seemed more like a busy hotel lobby than an emergency response command center. People were walking around and just chatting away, it didn't seem like anybody had an idea of what was going on. We found the Operations center and found by sheer chance Mr. Irwin. He was a short pudgy guy with reddish brown beard and hair with glasses. He was occupied with evacuation operations and directed us to the Director of Operations for the EOC (Emergency Operations Center).

Kyle and I received a vague briefing, and were told to wait for further orders from command. It was a room of mass chaos, people yelling and swearing at each other, and others running around like chickens with their heads cut off. Kyle and I found a command physician and got a briefing on emergency protocol orders. It was pretty much what we expected him to say, "Follow standard orders and do whatever you can to sustain life." It situations like these (mass casualty/multi trauma/multi incident) there isn't really time to sit and try to get a hold of a doctor to find out what to do especially if you know what needs to be done.

The command center did not have a lot of information for us, no phone numbers, no maps, no communications devices, no names or locations, to say it was disorganized would be an understatement. After the command center, we walked over to the logistics office and received paperwork to keep with us for records of our missions and our daily activities. We also received our vehicle deployment identification numbers 152 and 153. The command center, especially the operations director wouldn't have time to remember every ambulance service and every member of personnel that was here it would be too confusing and with the number system it could infinitive for how many personnel came and went. You have to remember, nobody knows how many people, apparatus, or other resources are going to are and you need something is flexible so as the operation grows so can the system.

We displayed our numbers in the front and rear windows of the trucks (ambulances) and then told to go get supplies for our units. We headed over to the supply warehouse and went over a list of things that we equipment we would need an abundance of. We definitely weren't short on food supplements with the water and the power bars that we had before we left Pittsburgh, but we needed to make sure we had plenty of medical equipment and supplies. We saw a lot of contamination (bio-hazard) protection equipment such as suits, gloves, and boots and each grabbed. We also grabbed a lot of medical scrubs and adult diapers for elderly patients and other patient who had no clothing or their clothes were contaminated. We had no idea what chemicals or other potentially harmful substances we were going to be around and didn't want to take the chance. We grabbed a lot of medical wound care patient treatment equipment such as trauma scissors, gauze, kling, band-aids, and almost every other thing you could possibly think of for every possible situation. We completed a quick inventory and headed back to the command post to wait for further orders.

Wednesday 8/31 1530

SEEING ALL THE confusion in the Command Center makes me wonder if these people have a clue of what they are doing. Waiting in the lobby of the local college front offices it like being in a fine art gallery. The floors were a dark brown stone tile polished with the main foyer having a fountain right in the middle of it all, made from stone and a smooth marble. The centerpiece went straight up like a chimney and had a religious engraving in just at eye level. The walls had painting of past religious figures or of public officials, with one in particular painting that I found to be rather humorous. It showed one public official painted in the center of the frame larger than anything else from the chest level up with smaller faces of an African American community. It almost made you think that he thought he was some sort of messiah or a savior to the African American community.

We finally got orders from a short older lady with curly red hair; we were to transport patients from one location to another. The area was an evacuation site field hospital to an out of area hospital. Before leaving we asked for area map books and directions, but they didn't have any. The lady just assumed the place we were going for our transports would have directions for us. 20 ambulances loaded up with a local ambulance taking point because none of us had a clue where anything was and EOC gave us clearance to go lights and sirens to the facility. The location was LSU campus and the field hospital was inside a dome building named the Pete Maranovich Athletic center known as the "P-MAC". Pulling into the parking lot was a nightmare as they had emergency and civilian vehicle scattered everywhere with no place to go. People were running around just getting in the way and some really didn't seem to care, they had no guidance outside and it was getting worse before it was getting better. We managed to park right outside the building and had a member of the event staff secure our truck so when we came out it would still be there. In the neighboring practice football field, the helicopters seem to keep coming and going continuously dropping off people and just lifting off to go get more wherever they were, the helicopters would find them.

Inside I saw a very well oiled machine of coordination by a group of Naval medical personnel. Doctors and nurses working as a unit despite overwhelming patient numbers, they were able to keep a system of triage flowing. It was truly an incredible sight to see through the crowds every patient was triaged, treated, and prepared for evacuation. The room was filled with a sense of organized chaos.

After loading two patients into our ambulance, one on the main stretcher and the second on a lightweight litter secured to our row seat (bench seat). Both our patients had mild medical issues, but they still required hospital treatment and admission. I was able to speak to a Navy doctor who gave us general description directions, but I had a feeling we were going to be on our own. They told us it would only take an hour in half but two hospitals and three hours later we arrived at our destination and I was more than happy to be able to get these to lovely ladies through the doors. The staff came outside with gurneys and was more than willing to help us with our patients.

We started back for the P-MAC and I couldn't help but shudder at what had finally occurred to me. All these people were going to keep coming, day and night, and it didn't show any signs of slowing down. We were going to have to work like robots, like a machine that wouldn't stop until the job was done and only GOD knew when that would be.

Thursday 9/01 0500

AFTER A LONG first day of non-stop chaos and transports it was finally good to get some sleep. EOC had to scamper to find cots after they realized they had no idea where they were going to put us up at night. The foyer/art gallery we were in earlier was now our big bunkroom with cots and tired personnel just trying to get some sleep. The cots were new and the material that was used to make the cots wasn't stretched so every time someone would move you could hear the material stretch, but it sounded like a fart, it was pretty funny. We crashed hard onto the bunks at 2330 (11pm) Jack and Gary were out cold as soon as their heads hit the pillow. I waited for Kyle so we could talk about what we were going to be doing tomorrow. I have no idea what in the hell we are going to do with all of these people, the hospitals are full and people keep coming in. It's going to be a long stressful time with no idea when it is all going to end, I just think about my family an how much I love and miss them.

The other people around us seem to have the same look of confusion and disbelief about all of this. It is one thing to see it on a television screen, but when you are engulfed by it, it becomes something more than just a movie. It's hard to sleep when you close your eyes and you still see everything and it stills to your mind like gum on a shoe. You have to work so much to get it away from you, but no matter how much you get out of your mind, there is still some always there. At 330am they woke the EOC woke us up to head back to the P-MAC for another group of patients that needed to be transported to all different hospitals around the area, after another four patient transports, we returned to the EOC and got back just in time to get ready for morning briefing.

It's hard to imagine there are places like the P-MAC all over the place, surrounded by pain, death, and suffering only to realize there aren't enough resources to save them all. There are those that are going to die as a result of what has transpired here, infection, drowning, heart and respitory failure, dehydration, exposure, and other environmental compromises. People will try to drink the water, or swim and walk, or they will try to survive without anything and die slowly.

The amounts of chemical and disease in this water has got to be frightening, not to mention the animals, the insects that carry infection, disease or bacteria that will be swarming from the stagnant water. This is going to be a health problem and epidemic for years.

Thursday 9/1 1200pm

SEEING ALL THE confusion in the command center I could tell it was going to be a bad day. Planning and execution can be a very difficult process to complete when the left hand doesn't know what the right hand is doing. EOC has all of these resources at their disposal and still can't get the ball rolling, there are close to 2 hundred ambulances and double the number in personnel and we are just sitting here. We got back to command post and as a group were all briefed on the plan of today's missions. We were told to load up on extra fluid supplies for triage centers that we would be going to that might be running short. Things like bottled water, IV fluids such as Saline (salt water) and Dextrose (diluted sugar) would help many people who were too weak or injured to be able to drink anything. They didn't have very much for pediatric and infant supplies, but there wasn't very much we could do about that anyway.

We had approx 13 ambulance loaded up and ready to go with additional personnel (a medical team of doctors and nurses) and supplies stacked from floor to ceiling, we were just waiting for final word from Command as they kept changing plans. We had a long drive of 63 miles from Baton Rouge to New Orleans and we didn't even know if there would fuel station for us to get back to the command center.

While we were waiting I finally found a radio station that was just seemed to take every worry or concern away, it was some morning radio show, I think they were out of Texas, but we got them in Baton Rouge, this group of radio personalities were absolutely hilarious and really helped to keep our minds off of why we were there to begin with.

Thursday 9/1 1251

DRIVING DOWN INTERSTATE 10 eastbound we got a clear view of what Katrina did here and you really can't help but kind of get that lump in your throat feeling. Trees snapped like matchsticks, houses ripped and destroyed from their foundations, even railroad tracks were strewn around like they were made of straw. We were actually driving over Lake Ponchitrain and you see the magnitude of how powerful the water was, as the waves and the current seem to have a life of their own slamming themselves up against the concrete pillars that hold this expressway together. It's kind of scary to drive directly over the accomplice to all of this death and destruction. We met the maker of all of this in Ohio on the way down and now we were seeing the contributing factors. You look at it from the driver's side window where you see the open water and think to yourself, "Hey its just water, what can it possibly hurt?" Then you look out the passenger side window and you see it's true power and possibility, the peace and tranquility of the water was destroyed by the overall picture of a region practically wiped off the map.

As we neared New Orleans in a town called Kenner we came to a triage center that used to be a highway under pass named Causeway Blvd. There were people everywhere it was just absolute chaos. One side of the highway had the survivors and refugees that had been evacuated either by boat or helicopter and the other side was the medical and evacuation center and all that separated us from the refugees was a metal guard rail and a handful of law enforcement personnel. Each person who came through was quickly triaged and medically cleared and were ready to be transported by bus to the airport where they would hop on a plane and shipped off somewhere else. The area of medical personnel had one triage center for critical patients, another triage center for people who had minor injuries and just needed an ambulance transport to the airport and a third one for those who just needed to be triaged and cleared. The smell of human feces and urine was everywhere, there were no portable toilets anywhere so people were going to the bathroom either behind bushes or wherever they were standing. The heat wasn't helping much as other smells of rotting food, stagnant water and the worst smell of all, death seemed to fill your nostrils with a sense of brutal reality. We were easily outnumbered and we knew it, we figured it was would only be a matter of time before we either got over run or additional law enforcement and or a military presence arrived.

The Operations Officer in charge of the Causeway told us (Bruce and I) to go down to a make shift boat dock about a mile down the road and assist the rescue crews bring people up to the evacuation site so we could keep those areas clear. Nothing hinders a rescue operation more when civilians who are getting in the way of completing your missions and surrounding you it just becomes a major problem and safety issue. We drove down the road as ordered and just as we came over a small hill where Interstates 10 and 610 intersect we saw the water right in front of us, it was like coming over the other side of a roller coaster just as you reached the top before sudden drop and there it is right there smacking you in the face. It was one of the most incredible things I have ever seen in my life. The water was easily 5-6 feet deep, roads were just gone, cars most of them completely submerged except for a glimpse of a roof or maybe the radio antenna

Looking around you could feel the impact of how devastating this was, the silence around us was the most chilling, there we were standing at the intersections of major highways and there was nothing, absolutely nothing. The ramp where the water met dry land was the makeshift "boat dock", with 10-12 boats sitting there tied to the rails of the on ramp. It's something really to behold as you think about all the people that have been rescued by the very boats that were sitting in front of me. The staging area was a 5 by 5 tent to provide cover for those waiting to leave. When we arrived there were people waiting for us, one guy had a couple of rifles. Let's just say that wasn't the comforting sight I had hoped for as we asked him to unload the weapons. Bruce loaded them up and took them to the Causeway as I stayed behind and waited for more people. Bruce came back a few minutes later as the rain fell mildly, a canoe arrived with a family and it was a gratifying feeling to see people safe and alive. The told us about how they had leave one of their loved ones at a near by road, just visible from where we were after the canoe had started to take on water. Bruce got the family down to the Causeway and continued to watch as helicopter after helicopter would fly over looking for others. 20 minutes later as I stood there on the water's edge with my binoculars, barely able to see the man next to a dump truck, a black hawk helicopter flew over and scooped up the man, let me just say that was truly a adrenaline rush to see that happen before my eyes and know I had a hand in that rescue. Bruce came back a short time later and told me how he rode on the black hawk to help the helicopter find him. I thought he was full of crap, but when it comes to situations like this, if you have knowledge about helping another person it's better just to take them along.

The last boat to come along was a couple of guys from a local power station who just wanted some water and food to take back with them. These guys were workers for the local sewage authority and were working day and night trying to get emergency power to their station so they could get the pipes to start flushing water out. It's amazing to see these guys sacrificing so much to try to save the city they live in, they weren't in boats or in an ambulance they were in a building trying to make a difference, to many people that dedication to a project like that may seem irrelevant, but to them they know what they are doing is important.

Thursday 9/1 2045

THE HEAT AND the humidity after the rain that was falling just seem to add to the nightmare we were living. It brought the smells of death, pain, and despair right into every part of you, like a film of filth that you can't wash off. All five senses of my body seem to detect the emotions of everyone around us. The frustration, anger, tears, blood, sweat, and fear ran off of not only us, but also the people we were helping. Fights were breaking out and it was over people just wanting to get away from this hell that they were living. The medical triages posts just couldn't handle the crowds despite having armed personnel there for security. People seem to come out of every direction and were becoming a growing concern; people needed medication that would help with medical conditions such as diabetes and asthma and we simply didn't have enough.

There have been times where it seemed like we were tripping over each other trying to get to people or supplies that were badly needed. From bandages to oxygen it was just total madness, you couldn't turn away from it there was no running or hiding from it all, there was no place you could go that didn't have someone sick or dying in front of you. We had three stations of triage and we needed law enforcement to keep everyone in line and from over taking the stations. It worked for a while until we ran out of buses and ambulances and more people kept coming in, then it got bad, there were fights breaking out, rage and panic filled every corner of the Causeway. People were being trampled as other began to throw things at us; we had to leave before someone got hurt.

Leaving the command post was something that we needed to do as the events that transpired had made us think what we were in for during the next several weeks. Too many patients and the environment became very unstable as the day went on. We were trying everything we could, but there just weren't enough of us. There were people with illnesses that made you sick to your stomach and pushed yourself to the breaking point of using every skill and resource you could possibly think of. I think of the patients who were sick and dying, who wanted to die rather continue to live and suffer through this nightmare. One gentleman tried to lie under a bus and kill himself because he lost his whole family. A mother with her teenage daughter who was so exhausted that the mother was having to drag her to us and all we could do was lay her down in the grass, give her some water and power bars to help her get her strength back. We couldn't take her with us; there was no place for her to go.

At nightfall we were ordered by the National Guard to shut down operations and get out of the medical triage centers as no more ambulances or buses were coming in until the morning. I couldn't believe what I was hearing as we hauled ass to get everything packed up and grabbed that last group of patients that we could. Bruce and I would be one of the last ambulances out of there to make sure all medical personnel got out safely. As were getting ready to leave, a big rig came by with sleeping cots and food and in that instant we could see the fear and desperation in the crowd of approximately 3,00 people. A woman, a mother of two was literally beating people with the very cots she had in her hand to keep them from stealing the cots she had gotten for her children, it's something that you just look at it disbelief as its right there in front of you not even 30 feet away. The military brought in additional personnel to help keep order so people wouldn't riot or try to kill each other.

We used every resource; every bit of strength, every skill and piece of equipment we could find and it still didn't feel like we accomplished a damn thing. We helped 30 and 300 more were waiting, it was just so hard to understand all of this, you look out into the world in front of you and you see all these people who didn't deserve to have to sleep under a highway over pass. Hell, didn't even describe this anymore, but there aren't any words that really can and yet having to experience all of this and try with every part of their being to get through it.

The drive back the EOC was long and tiring, as I just want to get back and crash on my cot, we caught up with Kyle and the other guys and grabbed some food so we could sit and debrief about the day's events. Kyle and his guys were traveling all over the state transporting critical patients as far as Lake Charles (4 hours one way) and looked just as exhausted as we did. EOC was starting to shut down operations as they were thinking about just performing operations in the day and keeping with a curfew so crews would be safe. We figured we would head to the new sleeping quarters they got us in the compound and get some sleep. The new quarters they had were much better than that small foyer, it was a wide open office building with two rooms that were bare empty, there wasn't even carpeting, just two large spaces with oscillating fans to keep us cool.

I don't know if I will be able to sleep tonight with the images I saw today, will my mind play tricks on me tonight or will I be at peace for a few hours? It's a question that is a crapshoot as I just try to close my eyes and hope for the best. The sights, sounds, and feelings of what we all have been through will haunt us for a long time, for how long no one knows. The only that is certain in my mind, with the scene we just left, instead of patients on stretchers, it will be bodies going into bags.

Friday 9/2 1029

MY ALARM ON my watch went off exactly at the same time it always did even when I was back home, 0645 (645am) and it became very obvious that is was one of those mornings it would be faster to just changes clothes rather than shower and just wear a stronger deodorant. It was a gut feeling in me telling me that there wasn't much time to do anything more than grab a bite to eat and get ready for the day, we all knew it was going to be a long one. EOC had an early morning all out mission briefing and they finally told us what was going on in a much larger scale, it was comforting to see EOC talking as a group and planning a course of action and getting all of its resource ready for the day. When there is a plan of action, you can see and feel the confidence not only in the command staff, but more importantly the field personnel as suddenly charged and motivated. After the briefing we all headed outside and all you could see from one end of the parking long to the other vertical and horizontal were ambulances and rescue vehicles, it looked like we were getting ready for war and I was ready to kick some ass. We were all ordered to load up our ambulances with, as many supplies as we could gather it didn't matter what it was and fill our ambulance completely. One by one we drove up to the loading docks and gathering whatever we could grab, it didn't matter to us, if it was in the warehouse of supplies we grabbed at least one of them. The teamwork being shown was amazing, the men and women helping others, race, ethnicity, religion, nationality none of that mattered we were all here and we were getting the job done. In one corner I saw a bunch of big bags of teddy bears and grabbed one of the bags and threw it into the ambulance, if there was a young child out there without a bear, they were going to get a bear.

A rough count of 250-300 ambulances with twice the amount of personnel had shown up wanting to help and it was most welcomed, as here we sat waiting for orders. It felt great to be part of something that wasn't about individuality or even the "what's in it for me" type people. We were all here for the same reason, because we needed to be. Our mission along with about 50 other ambulances in a long convoy of 200, were to go down to the Causeway and deliver supplies along with a medical team. We filled our ambulance according to the medical team's request and saddled up to leave. Suffice to say from the time of briefing to actual mission deployment it took almost 2 hours. EOC kept putting us on hold to make sure where we were all going. It was total chaos and frustrating. At one point we were going to the Causeway, then to the airport, and back to the Causeway. 50 ambulances headed down to the Causeway expecting the worst. If it were to be a repeat of last night it would be a very long day. We were given authorization to go lights and siren down there (we didn't have much choice) and we began the long 60 plus mile journey. Looking at the other motorists on the road I can only imagine the sight they are seeing with this long convoy of over 200 hundred ambulances traveling together with all of our lights on single file traveling down a road headed straight into hell. That is the only way to truly describe what you see when you get to the places we have been to.

You try not to think about it as you just push yourself forward despite what all your senses are telling you to do otherwise. The heat the sun, the smell or death, feces, urine, blood along with the sounds of children crying people screaming and yelling for help and the visions of it all that are forever etched into your mind. "Don't think about it, just keep going there will be time to fall apart later" that is what you tell yourself and others as you see them start to crack.

Friday 9/2 1248

T HE CAUSEWAY IS still a total war zone, there were bodies on the side of the road separated from the evacuation site that were waiting for body bags so they could be evacuated from the area. The smell of them just lying there even in the shaded area was enough to just make you puke whatever it was you ate earlier. By the time we had reached the Causeway over half of the convoy had split off of from the main group to go into downtown New Orleans for what was being called a massive evacuation operation. We didn't know what exactly was going on, but from what we were being told it was pretty bad and only getting worse.

There were people everywhere as it was almost impossible to tell the medical personnel from the civilians. Bruce and I stopped our trucks right outside the medical triage area and started to help load people into other vehicles when out of no where two guys grabbed our stretcher and took off into the triage area. It seemed like no matter how many we got out of there, there were more and more coming in, people were getting treated and released faster than the trucks were coming to get them. Our stretcher came back about ten minutes later with a young girl about 17 year old on it. She was a quadriplegic who was having seizures, so we loaded her into the ambulance when something happened that would forever change my life.

A rescue worker came running out of the crowd with a little girl in his arm and told me she was unresponsive with shallow breathing and she needed immediate medical attention. He put her into my arms and I told Bruce to haul ass to the local field hospital. This little girl was slightly ashen (which is bad seeing that she was African American) and was limp in my arms and warm to the touch. She was in bad shape and to make the situation worse, the family was in the back with me. I laid her on a little countertop in the back of the truck and immediately started giving her oxygen.

I couldn't cool her down too fast or she would go into shock and possibly make her heart stop, so I kicked the air conditioning up and kept her on the oxygen. Slowly but surely, she began to move her arm and legs, and then her eyes opened and I swear I thought she was the most beautiful little girl I had ever seen. By the time we reached the triage area she was doing better, but still very weak and tired.

The hospital was the Louis Armstrong Airport in Kenner about ten minutes from the Causeway and I was glad it was there. I told Bruce to open the rear doors and I would take the baby in myself with the family following close by, well that didn't exactly go as planned. The doors opened and Bruce grabbed the kid from me and headed inside without even knowing what was going on. It was one of those moments when I really didn't know what to think and ended up standing there wondering, "What in the hell do I do next?" I got to see her later on and it was great to see that through all of his this little girl was safe and she was going to be just fine.

We were ordered by the incident commander at the Causeway to stay at the airport once we go there so we secured our patients, and got to work in the triage centers. With our stethoscopes around our necks, a pocket full of medical gloves, and a triage clipboard Bruce and I started out on what we knew would be a long night. Performing triage duties is one of the most important jobs you can have when it comes to working a mass scene like this. Everything is set in motion off of your triage assessment for how your patients are treated medically and how fast they can get out of that particular area.

Walking around the airport we saw a living nightmare, there were people everywhere with all sorts of problems, men, women and children of all ages and race lying on the floor, in chairs, on luggage carriers, wheelchairs, etc it didn't seem to matter anything just to make comfortable. Some had little or no clothing and were protecting themselves from the elements with sheets and blankets. Once Bruce and I finished out triage assessments, we notified our area commander and started with our detailed treatments, we were on our own unless something serious happened. The medical personnel were stretched to the limit that no one could be spared unless they were absolutely needed.

We were doing everything we could for these people and it didn't seem to be enough, between dressing them in extra medical scrubs, to basic treatment it almost felt like there was no point to doing this, it seemed frustrating to look at all of this and not want to give up and just quit. You look into the eyes of these poor folks and see their pain and fear and it smacks you in the face and tells you to suck it up and get to work.

One patient I had was a sweet lady in her late 30's who weighed almost 400 pounds, she was in a lot of abdominal pain and found out she might have a urinary tract infection. We (me and a ambulance crew that had a nurse with them) had to wheel her on a luggage carrier to a desolate area so that they nurse could put a urinary catheter into her. I could see the tears of humiliation running down her face and all I could do is hold her hand as she went through this misery. Another woman was having a lot of trouble breathing so the ambulance crew and I ended having to intubate (put a tub down her throat) to help her be able to breathe. We did what we had to do to keep these people alive and safe, but I still felt sick to my stomach for having to witness or do the things we had to do.

We couldn't keep up with it all, the demand for us was more than we could handle, it was hell, pure hell and it was all around us. People were dead or dying, people were scared and angry, and we didn't have any answers to the questions they asked, all we could do was give them a assuring look that everything would be okay and they would be getting out of there soon. In EMS they always told us," be honest with your patient, don't lie or build false hopes to them." but how in hell do I look someone in the face and tell them they are not doing well and may die? Does that make me a better person than the guy next to you? I don't think anyone can answer that question without having a God complex or being a complete cold heartless jerk, or maybe it makes us stronger on the inside, being able to deal with the harsh realities of this job, that people do die.

Treating while waiting for evacuation *P. Trotta*

A beautiful scene of New Orleans camouflaging destruction and death

A child's stuffed animal lost in the madness

A soldier holding a little girl at Louis Armstrong airport

Abandoned car, the wave rolled the car over as we drove by

An icon of hope and faith that was left behind

Driving lake Ponchitrain, the accessory of Katrina's wrath

Highway off ramp (2 bodies were found near ramp in 4 feet of water)

Evacuees waiting to be treated at Louis Armstrong Airport. *P. Trotta*

The Green Team at Louis Armstrong Airport *P. Trotta*

Surveying the damage of the convention center. *P. Trotta*

New Orleans SWAT (Special Weapons and Tactics) at convention center

We even rescued the four-legged victims

Refugee who refused to leave, found dead 24 hours later

A clear sign that the flood waters were receding in St Bernard Parish

Trash and debris floated past this sign (Unlawful to Litter) as we drove by

Personnel and supplies after the Causeway evacuation *P. Trotta*

A boat crew performing another search for victims/survivors

A rescue helicopters awaiting patients, the real heroes

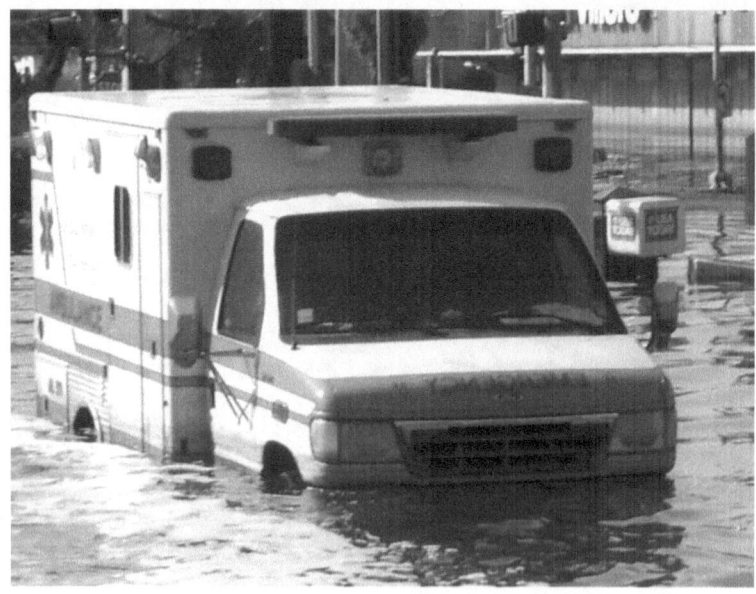

Abandoned ambulance in 9th ward surrounded by 3 feet of water

Rescue aircraft landing at Louis Armstrong Airport

Cases of water, pool of ice, and oxygen for those in need

Emergency Operations Center (EOC) Baton Rouge, LA

Hotel rooms blown out by heavy winds

People grateful to be alive after Katrina *P. Trotta*

Aftermath of the airport evacuation center *P. Trotta*

Fueling for the 60 plus mile trip to New Orleans

Patients being unloaded from luggage carriers as helicopters wait

Rear wall of storage unit collapse

Residential neighborhood in 3 feet of water

Friday 9/2 2028

THE AIRPORT WAS a disaster and it just seemed to get worse, I was doing everything from blood pressures to helping people make MRE's (Meals Ready to Eat military food packets). People who had arthritis or couldn't see without their glasses couldn't even open the packets of food they were given. For three hours I did nothing but prepare these meals so they would be able to eat something, as many of them hadn't eaten for hours some even days. People were lined up outside waiting for help and inside waiting to leave for hours on end, it was an endless ocean of people and all I could do was go one by one trying get each one medically cleared and out to a plane. My body was tired and achy and mentally exhausted to the point where I was on autopilot, just going and going not wanting to stop.

I went outside for some air and a truck driver came up to me and told me something that made me want to just drop over. He and two other drivers had shown up with people in the back of the trucks needing medical attention. I told him to pull up to the front doors and I would get help. I ran inside and remember just yelling, "I got people out here and they need help." I swear I saw people coming out of nowhere coming to help, I had one group start moving boxes of material and supplies so we would have a place to put them while a larger group followed me outside to get the people out of the trucks. Looking into the first truck you saw something that in all my years I had never seen before. These folks were strapped to anything and everything like backboards, cubicle walls, sheets of wood, table tops, doors with the knobs taken off, you name it they used it, it was incredible. I didn't ask where they came from, it didn't matter we just needed to get them out of thee trucks and into the triage centers.

We hopped into the trucks and one by one we worked a system of perfection with men and women who came from everywhere ready to lend a hand. It was a like a bucket brigade of handing off patients to people ready to take them inside. In groups of four, the patients were handed off and carried inside and secured, once the patient was secured, the group would come back out and be ready for the next patient. All three trucks were emptied and all patients were secured in the triage centers in less than 30 minutes, I have to say it was very impressive to see total strangers work as organized as we all were. Each member of each group took one patient and performed a triage and assessment having the whole unit treated within 20 minutes of their arrival at the triage center, it was the visual definition of EMS, you take what you have and you make it work. They were immediately taken off of whatever they were strapped to and placed into cots and stretchers with new clothes and food in the systems. You could see the initial look of fear and confusion on their face when they first arrived in those trucks, but once they were inside and treated they were smiling and the look in their eyes was of gratitude. You couldn't help but feel better about yourself and the job you love to do, as small accomplishments seemed to matter so much more than anything else.

These folks went from having lost everything and being carried off to some damn place where they didn't know what was going to happen to them next to now having food in their stomachs, clean clothes on their backs, and a cot to sleep on. You do your best to give them a little luxury and comfort when there is nothing else to give. To be able to sleep while waiting for someone to put you on a plane to take you to a place that has no flood water, no destruction and maybe a place that they can call home and start their lives over again. It's like people would say in desperate times, when you have hit the bottom there is only one place you can go, how fast you get back to that spot is up to you.

The helicopters would continue to bring people in as the planes would slowly take people out, in the middle of it all would be us, feverishly and continuously doing the best to make sure they got what they needed, that was what we were there to do.

Saturday 9/3 0837

WOKE UP AT 0530 (530am) and felt like crap, just can't get enough sleep my mind keeps racing about the days that we have had and the days to come and what will be in store for us next. I know as soon as I get home I am having a home cooked meal, spending time with my family, and going to enjoy the comfort of my bed mattress. I haven't felt this exhausted in years, but it's also a feeling that lets me know that I am working hard and doing my best. Breakfast this morning was the same biscuits and sausage gravy, I think they save the old ones and just reheat them because they could be used as hockey pucks, but hey it's free its here, and I have eaten a lot worse including nothing at all.

This mornings briefing by EOC has us taking another medical team down to the Causeway for what we are being told more people are still down there. There are reports of critical patients down there with severe cases of dehydration and heat exhaustion borderline heat stroke. The heat and humidity is inhumane and show no signs of cooling. I guess this truly is hell on earth and you can't escape it. We are taking a 7-man/woman team with us, 2 doctors 1 from Massachusetts by the name of Gregg Wolfe and the other was Stuart LeBas from Shreveport, La. The 5 nurses, Paula, Lisa, Sherry, Tanya, and Bob are a group of nurses from a hospital in Arkansas. Being around these folks take time from their own lives on their own time to come down and volunteer their services really shows the human spirit of wanting to help. Being around them and many others has been the motivational butt kicking that Bruce and I need to keep going.

Under orders by EOC Bruce and I loaded up the ambulance with supplies that we were told they needed down at the Causeway and then be in the convoy ready to leave on a moments notice. We lined up with the convoy of 30 ambulances all heading down to the Causeway and what was to be a moment's notice ended up being an hour delay. Just thinking about going back there makes my stomach turn as I really hate being down there, the heat and smell of dead bodies really seem to over take over anything else around you. Bruce and I gave the medical team a quick briefing on what we had been dealing with over the past several days, not just the Causeway but also the airport and from what we heard of other areas where EMS crew were working. They really seemed to listen intently and the usual shocked look on their faces was becoming to look all too familiar.

The convoy once again broke off at the Interstate 10 and Kenner interchange and I mentally prepared myself for the same sights as before down at the Causeway, but as we arrived what we all saw was a blessing, it was empty! There were maybe 30 people still there, but they were being treated and ready for transport to hospitals in Lake Charles. Speaking to the incident commander at the site, we were told that in the middle of the night 100 or so buses arrived and took everyone who was able to walk and medically cleared out to the airport to be flown out. Finally some good news to know that positive results were physically in front of us, unfortunately there were some (I never asked how many) that didn't survive being down there and they were taken away for the morgue. "If only we had gotten to them sooner, If only we had been faster, if only, if only . . ." these are the words I would try to keep out of my head. "There is no time to beat yourself up, there is still a lot of work to do and a lot of other lives that need to be saved. Make up for the loss you couldn't help by saving others and making it home yourself." We repeat that over and over and then we would get back into our truck and we go to the next place and do our best.

We told the medical team about what we had experienced yesterday at the airport and got clearance to head there, the Causeway was done and we were looking to help so figured to go there and see what could be done. We arrived and Bruce and I found it exactly as we had left it the day before. They had gotten some people out, but there more were still there with many more coming in. Dr. LeBas told us to unload the supplies from the ambulance and secure in an empty spot and to set up a station as Dr. LeBas went to go find someone in charge. Dr. LeBas came back with a couple of tables and I went for specific supplies he asked for. Instead of a general triage center, Dr. LeBas came with an idea to have a section set up for specific medical treatment for specific medical ailments. That way we could focus on a few specific things and keep the station controlled as much as possible. Each patient would be triaged with the basics first and then the specific treatments would be done to ensure that the treatment was working. We grabbed glucometers (they measure blood sugar in diabetic patients) and insulin for anyone who needed it for diabetic ailments.

We were each given assignments; Dr. LeBas would be the station coordinator, Dr. Wolfe would be the main physician (writing prescriptions and prioritizing patient needs), the nurses would perform advanced care measures (IV access, intubations or anything else) and Bruce and I would perform basic triage and supply acquisition. It was the perfect system that would have all aspects of patient care and coordination performed quickly and effectively. Our unit would be called the D & D unit (Diabetes and Dialysis) for patients with kidney/renal compromise and diabetes related illnesses.

Throughout the day we treated each and every patient that came into our section and it didn't matter who they were, we each did our part and many people were treated and helped. We did it all, fed them, clothed them, gave them medication, but most of all we gave them piece of mind. That was the most important thing to our patients was just them knowing that they were being taken care of and we were there for them. Some needed oxygen to help them breathe, some needed insulin for their diabetes, but all of them just wanted some reassurance. All of them asked the same question just wanting to know what was going to happen to them, it was hard not to be able to give them any real answers other than, "better than where you are right now". In some crowded airport terminal, surrounded by people who were just as scared and angry that just wanted answers, they didn't want lies or promises just the truth.

The most aggravating thing out of the whole thing had to be the constant annoyance of the press media. From all over the world, all these people cared about was getting a news story and not helping the very people they were interviewing. I was holding one ladies hand and letting her know she would be getting on a plane soon when this German guy comes up and shoves a microphone right in my face and asks me what I was saying to her. Well this scared the hell out of this nice little old lady and it only made things worse for her because she thought she was going to die. It took everything I had not to punch this guy in the face, but I decided just to smile and go with the program and maybe he would leave me alone. I figured if the public could hear me and understand what these people were going through maybe it could get some people to want to lend a hand. Money, clothes, food, even volunteering would make a difference to anyone who was going through this. This wasn't about headlines, fame, or anything else it was about the people of the Gulf Coast.

The unit filled up pretty quick and we ended up having to add one people with general illnesses. It didn't stop us from what we were doing, and it didn't stop us from making progress but we still needed to get our patients out the door. Many of them had been there from the start of our day and it was coming on 5 hours later and they were still lying there. One of the big guys from FEMA came over and I tried to let him know that we needed to get our people out on those planes, but he just blew me off. Right about that time though Dr. Wolfe got word that the FEMA guy wasn't going to help us and suddenly I saw a whole new side of the good Dr. Wolfe. Dr. Wolfe a guy who is about 5"8 160 pounds got up into the face of this 6"3 250 pound FEMA guy and told him that our people would be next to get on the plane. It was amazing to see Dr. Wolfe dealing with guy, it was even more amazing to see that it worked as 10 minutes later a colonel from the National Guard came up to Dr. LeBas and told him that he was given specific orders to take everyone from our triage area out to the planes.

Once everyone was gone, Dr. LeBas was given authorization to leave, the last of the planes were coming in to pick up the last of the patients for the day and the National Guard had control of the situation. Looking at our triage area empty and quiet, it felt really good to know we had made a difference to about 50 people and they were now going to be in a safe place to get their lives back on track. We loaded up what was left of our supplies (very little I might add) and headed back to the command center.

Sunday 9/4 0805

LAST NIGHT WE were able to have night to relax and take to an evening off, EOC shut down operations and told everyone to take a break but to ready early in the morning. Dr. LeBas treated the whole medical team to some dinner and fun at the local Hooters. Dr. LeBas paid for everything and just wanted to say thank you for coming from our own lives to help save the state he lived in. He bought us a round of drinks to give a toast to us, the team, and to the victims of this tragic event. We had wings, fries, and onion rings and just enjoyed each other's company. Dr. Wolfe was laughing so hard he kept falling off of his bar stool and he even spit an onion ring across the table. The food was great and the memories we shared were exactly what we needed to help us deal with what we had just been through. In this job, you need to find a way to talk about the bad days, the bad calls that make you cry, and most of the times you wonder why you do this job. Whether you laugh or you cry, as long as you let it out it gives the out that you need so you don't self-destruct. As a joke Dr. LeBas decide to buy Bruce and Paula a couple of Hooter's employee shirts and what had to be one of the scariest things I have ever seen was Bruce actually fit into it. That's a mental image that never leaves your head and just makes you want to wash your eyes out with acid.

Sunday 9/4 0922

THROUGH ALL OF yesterday I couldn't help but think about everybody we helped, you can't help but feel a little bit better about yourself knowing that for the first time in many years that you truly made a difference. You take a situation of overwhelming odds and horrible circumstances and make something positive from it. You wake up, get dressed, grab a bite to eat and saddle up for the next mission.

Driving down the interstate I couldn't help but think about the patients from yesterday, this elderly lady that had been married for 53 years this sweet lady lost her husband in the flood. Elliot the calmest nicest guy in the world guy in the world despite having no clothes of his own, we ended putting medical scrubs on him just so he would feel more comfortable. The most memorable guy was Willie, he was a character all his own and a truly nice guy who had a smile even though he had lost everything. Overall despite all confusion, anger, pain, and the folks we took care of never showed anything but gratitude towards us. There were a few difficult cases of one guy who had to be physically restrained because he kept pulling out his IV's, and then another lady who didn't speak any English who kept trying to take her clothes off. The one that hit me the most was an African American lady who had an open sore (decubitus ulcer) on her leg. I had to look into her leg and found it to have maggots crawling inside of it. I had to use a 5cc syringe with beta dine cleaner and insert it into her sore so the wound could be cleaned from the maggots and other infections. I quickly dressed the wound with a bandage and kling then went outside with thoughts of throwing up, it was one of the nastiest things I had ever seen.

I was ready to leave after that I knew that I had stuck through most of it all, but that was my breaking point and thinking about made it obvious that she was probably going to lose the leg. The infection and necrotic (dead tissue) was just killing her leg and was causing her a lot of pain. Through all of this it wasn't about just one person, it was about us all, Dr. LeBas, Dr. Wolfe, Paula, Sherry, Tanya, Lisa, and Bob for all their dedication and sacrifice. They were the true meaning of the phrase patient care advocates, they refuse to give up on the patient's they treated and refused to let each other give up. A wise man once said," Through each man's belief that courage is life, his strength will come when his passion for life is his one true belief." Through blood, urine, feces, tears, and all the rage we came through it and grew as people, a team, and most of all a new view of love for this profession we call Healthcare Providers and lifesavers.

Even through victory and gratification the work is far from over as we were ordered to take the medical team back down to the airport to make sure everything was handled. We got there and it was another blessing the whole terminal was empty, nothing was there, no patients, no triage station, nothing it was like God came through and took it all away. It was truly hard to believe that just 12 hours earlier you couldn't see the floor past the feet of the patients we were treating. Just as we were leaving, Donald Rumsfeld decided to make his grand entrance and survey our work; well I think he was a day late and dollar short. He came through and did his little walk through, but I didn't think too much of his appearance and probably neither did he.

We headed back to the command post and EOC needed us to head down to Downtown New Orleans to assist in rescue operations and patient assessment and treatment. The Green team, as they were called now were, going to be heading back home after a job well done. They had earned it, but now Bruce and I would stay behind and finish what we had started. I wanted to head home too, but our work wasn't done yet. Now that the outer areas had been cleared out, we (Bruce and I) were finally going to go into the belly of the beast and hopefully bring out some survivors.

Sunday 9/4 1243

DOWNTOWN NEW ORLEANS is in ruins; the city is completely destroyed with a trail of debris and shattered lives left behind. Driving down through the area we got to see the full impact of everything and what we were dealing with. Hotels with all of their windows blown out with the furniture hanging outside the windows or on the ground, trees that were literally catapulted through houses, and skyscrapers that had their rooftops blown off and found three blocks away. The Convention Center looked more like something of a bad movie rather than the sight of refuge and sanctuary. Driving down the main road past it, it was just mounds and mounds of clothing, food, personal items and everything else you could possibly imagine that were left behind. We had to move furniture, shopping carts, mattresses and many other objects out of the way just so we could get our vehicles through the roadway.

Damage and vandalism were just some of the sights and smells that seemed to come from every direction. Outside the convention center as we were driving by everything about the tragedy here was in front of us. A New Orleans police cruiser was parked in the grass and the car had been stripped of everything, from the light bar to the driver's side door was gone. I can't imagine what someone would want with a car door, but someone somewhere had it. There were walkers and prosthetic limbs left on the sidewalk, as puddles of dried blood from the intense heat and humidity seemed to be everywhere it looked like a battle between good versus evil and no one won. They battled hunger, heat, dehydration, environmental elements and each other just for a chance to be rescued. Those who survived showed true courage, will and spirit despite a system that failed to provide for them. The survivors of this will be the true heroes, not us they beat the odds when others fell and now they can stand a little taller and walk with pride. It's always been about survival and coming out of a bad situation to be able to tell the story of the event to prevent such travesties from ever happening again.

Those folks were abandoned there after they were promised that someone would be there to guide them to safety, but instead they found nothing but lies and a sudden sense of loss. They had to fend for themselves as no signs of help was anywhere to be found. Local civil services weren't able to do anything as they were having their own problems. Cars were underwater, buildings were inaccessible, and many of the personnel themselves were unavailable. I ask myself the question, "Who rescues the rescuer when they become the victim?" Sometimes the answers come slow when you need them quick and it sucks.

Seeing all of the food wrappers and empty water bottles you wonder how many people died before these crucial essentials arrived? Were those who died before help arrived better off dead than to continue praying for something that may not have come at all? They probably suffered a lot less with their last breath than those who were just trying to catch their breath. At first count 5,000 people were dropped off here with no food, water, or medical care for the first 24 hours and that was the worst. People were fighting left and right as others died in front of their own families, some people were reported to have been crushed to death while trying to sleep under a loading dock ramp. The body count, it'll be too high to even begin to think about, but the rumors alone are enough to leave these images as a horror in my eyes. The count is speculated at 800, but with the body recovery not yet beginning we all expect it to get a lot bigger and to reach the thousands.

Soon there will be nothing but a body recovery to be done and that is when we will know if we succeeded in our efforts or failed in a mission that seemed doomed from the start. If we don't learn from our past and our mistakes how do we prepare for our future? Through time we have always thought we could handle anything and it brought complacency and arrogance to our society. Through this way of thinking we have always allowed our guard to be down and our enemies to use it against us and then we have to spend years trying to figure out what we did wrong and now this was no different. Someday maybe we will learn, but the cost of that lesson will never be worth the lives that are lost.

Monday 9/5 1146

THIS MORNING WE were suppose to go and begin search and rescue/ recovery, but unfortunately both trucks were broken so we all had to stay behind to have them repaired. I guess it's kind of a blessing in disguise that way we can rest up and be ready for other missions that will be more and more difficult. I guess the problems with the trucks are from the National Guard fuel tanker trucks that we had been using for fuel back and forth from New Orleans to the command center. The fuel pumps didn't have filters on their pumps allowing all sorts of debris and residue that was in the tanks to go into our fuel system causing them to stall. It's amazing how much you miss a fuel station when you don't have one at every corner. Driving one way to New Orleans is about 65 miles one way and about 3 quarters of our fuel tank, so needing fuel to get back was pretty much a no option requirement.

Looking and talking to the guys, we all seem to be doing pretty well talking about things we have seen and done helps us keep focus on the job and not allowing anyone to fall apart or freeze up when they are needed most. Being broken down for the day has been a blessing in disguise, I think the power bars that I had been eating over the past few days had clogged up my system and now I had really bad cramps and diarrhea, I am talking serious pain and discomfort here, I think I finally got my system flushed out thanks mostly to downing a whole bottle of Maalox and multiple trips to the bathroom.

Walking around the EOC compound you really got a sense of teamwork and family, as crews leave and new ones arrive it's a sense of relief and accomplishment knowing that it will soon be our turn to do the same. The personnel continue to come in some from across the country some even across the ocean to lend a hand. I think about the folks like Becky and Mark from Emerson ambulance and Jaime and Julie from Jennings fire department who came in and did the best they could and have a lot of pride for what they went through. I like to think though that soon there will be nothing left to do as more and more people leave the city and begin their new lives elsewhere.

The streets of New Orleans can be some of the quietest around as you stop and look around at the homes that are empty as they sit there waiting for the families to occupy them once more. Will New Orleans ever be the same? Probably not, everybody and everything will have to start from the beginning, as many have nothing to start over with. Will New Orleans ever recover? If the human spirit is strong enough, there is nothing that cannot be overcome or overtaken. It will be a long and frustrating battle that will seem to have no end. You can always tear down an old building and put up a new one, but you can't do the same thing with a human person. A person has to want it bad enough and not let anything get in their way, the people of New Orleans the ones who stayed behind and braved the elements for what they love have that will and have the ability to bring New Orleans back to where it was before.

Monday 9/6 0946

WE WOKE UP this morning and were told we would be going downtown New Orleans to have a local paramedic ride along with us and run city service calls. Things suddenly were changed and then it became a plan of running search and rescue in boats, but apparently they didn't have enough boats to do the patrols they wanted. There was a boat crew who I was talking to who told me that they were leaving because of the constant passing around by FEMA, EOC, and the Fish, Game and Wildlife Commission. They would report at 0400, but wouldn't be allowed to put their boats in the water until 12 noon. They considered it a complete waste of time and decided to pack up and head home. There was a rumor that 20 boats crews, took their boats and bailed because they were also tired of the games from all the different agencies.

Sometimes I have thought we should have done the same, there was no communication between the organizations pretty much the classic phrase too many chief and not enough Indians. No straight or truthful answers, just a whole bunch of political agendas thrown in with a bunch of big words meant to confuse us. The reports they received were always incomplete, bad, or no reports at all. At one point there were several refugees brought to the command post for a few days and they ended up acting like EOC staff and made promises to crews that they had no business making. Kyle and I were including in the scam as the imposters told us to report at 0400 (4am) for boat crew assignments. We woke up at 4am and reported for duty and the EOC just looked at us like we were insane. We explained to them what we were told by the other people and they explained to us how we weren't the only ones that they made promises to. They had been escorted off the premises just two hours prior to us waking up for duty. It just made me wonder how many other personnel were screwed with and lied to by these people that shouldn't have even been at the command center to begin with.

We received updated orders at 1300 we finally arrived in downtown area at a staging area outside the convention center. It was a two-part parking lot with one part the staging area for us and the second part was a triage area and landing zone for helicopters.

The whole area was controlled by the National Guard and roving patrols of the New Orleans Police (the ones that remained) and the elite New Orleans SWAT teams. These guys were the best of the best at what they did and they backed it up, with not only performance but also mainly through reputation. At our staging area we were under the command of Ken Bouvier, the Director of Emergency Medical Operations and Disaster Management as well as Supervisor for New Orleans EMS. His reputation and methods of getting the job done definitely preceded him. This guy was old school, he didn't believe in politics of what we did, he just wanted to make sure everyone did their job and went home at the end of the day and his heavy Cajun accent just brought a level of respect that was pretty much automatic.

We staged 11 ambulances and were debriefed by Ken about what the plan was going to be and it involved the use of the National Guard. We ended up waiting for 2 hours for the National Guard to arrive with high water vehicles. Driving through the city you really got a feel and first hand glimpse of the damage that Katrina did. In many ways it was like a ghost town with no signs of any kind of life. We would come across people who refused to leave and just wanted food and water; there were others who were simply wading through the water as if it were nothing. There were all sorts of bacteria, chemicals, disease and God only knows what else in this water and people didn't care.

We even saw animals such as snakes, lizards, and even reports of seeing alligators, even after the waters would recede all the disease and other hazardous materials would remain for years. Some folks were living in makeshift shelters made from anything they could find, we came across was a husband and wife along with their two dogs living on a center median underneath a highway overpass. He refused to leave and just wanted batteries for his portable television. I couldn't believe this guy, he didn't want the help of personnel coming in and willing to take him to a safer place, not even food or water or medical attention.

Too many of the residents within this city were about two things: their love of New Orleans and not wanting to give up what little they may have had. For many their homes and their possessions were all they had and just didn't want to leave without them and as a result many people lost their lives and it seems so sad. The question that continues to bug me, is a home worth a life? Why do we as people sacrifice so much of ourselves for what can become so trivial and yet through natural instinct it makes perfect sense in a later time.

As we continue to travel through the city you could see the markings on the homes each one identifying which ones had bodies and which ones didn't. Each house had a large X, the number of bodies (if there were none there was a zero) and the date that it was searched. It was the true sense of relief to see many of the houses having zeros next to the X's. Whenever we would pass a house with a number, you couldn't help but close your eyes and say a prayer, the thought of someone's family member and or loved one had passed away in that home. To know that the family would have to wait for the day when they would be able to return to their home and lay their loved one to rest. Some homes had as many as 6 next to the X and you just wanted to yell out in anger and hope those lost souls didn't suffer. These were entire families were now gone, parents, children, husbands, wives it didn't matter who it was they were dead. It really makes you question your own spirituality and morality, you go numb and shut it out telling yourself there will be time for the pain and tears to come out, but now wasn't the time. In this job it's about saving those who are alive and leave the dead behind, there is nothing you can do for them except for say a prayer and move one.

9/6 2001

HEADING BACK FROM downtown New Orleans after getting to do some search and rescue in the high water rescue trucks with the Army National Guard. Climbing into the back of this heap of metal with wheels, Brock and I were greeted by two members of the I.C.E. (Immigration and Customs Enforcement) teams. Jeff who was from San Antonio, Texas carried a 12-gauge pump action shotgun; his partner named Oscar from El Paso, Texas carried a military issue M-4. Just before we went under way Oscar asked me if I knew how to operate the weapon he had with him. When I answered yes, he smiled and said if he was hit by gunfire he would expect me to pick it up and fire, suddenly I realized it wasn't just a matter of survival for the victims of Katrina but it was also a matter of survival for us.

Our specific orders, to try to save the living and under no circumstances were we to retrieve the deceased, it would have been difficult to get anyone alive to climb on board a truck with two armed men and a dead body on board. We came across men and women who refused to leave, despite the fact that the environment was unstable and the water continued to be dangerous and toxic. All many of them wanted were bottles of water as we noticed stacked walls of MRE's all around the city. At first I thought it was stupidity and pride, but to many of them it is all they own and just wanted to protect it from people who are robbing and looting homes as people are being rescued. Others said they just love New Orleans too much to leave it behind and all I can think of is New Orleans is going to get them killed. We came across a bridge that had a man who was completely unresponsive, my first thought was he was dead, but something told me to check anyways. He was sleeping on an old mattress cushions and using an old garbage bag for a roof. I checked him and when he woke up he asked me what my problem was. I told him I was from search and rescue and we were here to get him out if he wanted to leave. He told me no and to just leave him some water, I had to tell him that if he were to stay where he was he might die (he was pale with shortness of breath). Despite my best efforts to convince him it was for his own safety he told me he was staying and even gave me the middle finger as I was getting ready to leave.

As we drove through the city, you could see how different everything was, we would go one block and find no water, and no damage other than trash and trees down, only to drive one block down and find a foot of water. At one point the water got to be about 3-4 feet deep and found several bodies floating along the water's edge. They had been found by two search and rescue teams and were covered with blue tarps to protect their identity, but seeing the condition of the bodies, concealing identities would not be a problem. Their bodies were bloated and swollen, caused by the gases building up in their systems from the longterm exposure to the water and other elements. We came across bodies that were floating outside and inside of buildings; I saw one wearing a life vest that looked like he couldn't have been more than about 15 years old. You could see them in front of you just a few feet away, wanting to grab them, but knowing you couldn't because there were still many people alive and struggling to survive. Seeing the lives lost you couldn't help, but say a prayer for them, their lives washed away by such a senseless tragedy was something terrible, it really puts life into perspective. The smell of the dead, the waste, and everything else around was just nauseating, you tried to burn the smell out of your head by focusing on something else but its strong like a cup of hard coffee. The water would turn from green to black just by the turning wheels of the truck as we continued down the road. Cars that were flooded to the roofs would actually roll over onto their roofs from the water was so high it lifted the cars off the ground.

9/7 2022

WITH THE CITY of New Orleans in chaos getting the chance to work with New Orleans EMS and responding to their calls wasn't exactly easy. The city was still under curfew and that meant after the sun is gone nothing more is done inside the city limits. Any emergency calls that came through their 911-communication center would have to wait until morning to be answered by personnel. Each ambulance was assigned a paramedic from New Orleans EMS to help us navigate through the city so we would have an idea of where we were going and how to get there.

You had to give New Orleans EMS and the other departments a lot of credit for the sacrifices they have made by staying behind and continuing their work despite all of this. These folks had lost everything and yet they still showed up for work. If that is not dedication to not only the job you do, but also the city you live in, I don't know what is. Many of them were offered vacation packages to get away from everything and just relax and even with that offer many of them stayed and continue to work. I can find no fault in those who left, they have been through hell and deserve it, but you want to shake all of their hands for the work they have done and everything they have had to endure.

Our medic Camille was really cool and seemed to very laid back as well as quite popular with the locals. Driving down Bourbon St. everyone seemed to know her as they yelled her name and waved to her, at times she seemed more like a tour guide pointing out buildings and particular streets that were significant to New Orleans. We received a call for a person down and were also given specific orders by EOC and city coordinators to make entry into any house or structure by any means necessary.

We checked the house and found no way in so I had to kick the door in and let me tell you how much of a pain in the ass that was. After three good kicks the door along with the dead bolt finally gave way and we made entry. Having nothing other than my 7-inch knife I had no other weapon with me so making myself known and announced was a critical need. We found nothing and began to leave when a neighbor came by and asked us who we were. We told him who we were and found out they called 911 to have a body removed from across the street. To do something like that after not having done for several years was a little bit of a rush as you go into a house you know nothing about with no knowledge of what could be on the other side of the door.

Driving through the French Quarter, the spirit of Mardi Gras was very enticing, especially to see it in person when the only other time you do see it is on episodes of COPS. Looking around it was amazing how intact everything was, very little structural damage like trees and general garbage and no a single drop of water anywhere. There were bars that were actually open and operating with people inside having a good time drinking and laughing as if it were just another day. I guess for some the party never really ends even when it looks like the world just did, no amount of destruction, wind, rain, thunder or flooding can stop some from wanting a good cold stiff drink.

We were later called to a bar/strip club in the French Quarter for someone who called but kept hanging up. Just as we arrived a second ambulance arrived and found once again no visible entry, myself and two other medics found windows open on the second floor and actually started climbing up the neighboring building to reach the windows. When I got inside the building, I came into an apartment that had multiple weapons mostly swords inside of it. Not knowing if anyone was home, I announced who I was because the last thing I needed was some crazy person wielding a sword at me. Doing a quick search of the apartment, I found ten samurai swords and multiple knives and thought it best to get the hell out of the apartment before I got myself into something I wouldn't be able to get out of. Checking the rest of the building we finally found the woman who needed help. Scared and unsure of what to do we told her if she wanted to leave she should grab a few small things and we would take her to an evacuation site.

Standing outside the building we got ready to take her in our ambulance and some of her friends showed up. They tried to talk her out of leaving and actually got ignorant with us to the point where law enforcement had to tell them to walk away or get arrested. One of the guys was the one who had the apartment full of swords and advised him to secure them better or the police would confiscate them. Taking her to evacuation site you could see she was scared, she had run out of food and water and couldn't take it anymore. She told me she had been doing a lot of drinking and even smoking of marijuana. She thought she was going to get arrested because of the drugs, but I had to assure her I wasn't a cop and what she had done didn't matter and she would be taken somewhere safe.

We came across all walks of life people who had dogs who refused to leave until they were picked up by the SPCA (Society for Prevention of Cruelty to Animals), we even saved a stray dog and being a dog lover myself it made me feel good about myself. Through it all over a span of 4 hours I felt like we had truly made a difference, getting 5 people to safety and rescuing 3 dogs it was something that I will always remember. It's something that you can take with you for the rest of your life knowing that despite all of this wrong, you did something right and now these people will have a better chance than they did before.

Getting back to the staging area (the same one as when we were with the National Guard) one of the other crews said they thought they saw a body outside the staging area in some bushes near by. Out of pure morbid curiosity a group of us decided to walk over and see if there was anything we could do. There was a body in the bushes, but he wasn't dead as we noticed he was breathing and when we touched him he moved and actually started talking. After getting over the initial shock and making sure my heart was still in my chest we helped out of the bushes and walk him over to the staging area. He told us he had been lying there for three days without food or water and was hiding from the cops because he didn't want to get arrested. Sitting in a chair this guy drank two bottles of water, a bottle of orange juice, and cleaned out a complete MRE in a matter of 10 minutes. He told us he was having seizures and didn't have his medication after his house was wiped out because of the floods. When we told him we were going to take him to the triage area he got a little aggravated He really didn't put up too much of an argument when we told him if he didn't go he would be arrested, after that he was very cooperative and was willing to go. It felt really good to see the triage area having people that were willing and wanting to leave many of them were the very people we had rescued. If he had been there for three days like he said my question was, why hadn't anyone noticed him there before? Didn't anyone care or was it just through all the confusion and chaos that no one looked around to see if anyone was there? The Convention Center was the site of so much death maybe in some way everyone just gave up hope that anyone would ever save them. Was this a victory in a battle that surrounded defeat? Some might say, but when I see the faces of those who were now safe, it should me that thee was a positive fulfillment to this job.

9/8 1045

TO START THE morning I woke up saying a prayer to my Dad, Phillip who 8 years ago today died from a heart attack. Even to this day I miss him very much, but I know he is always watching over me and making sure I don't do anything stupid or get myself killed. He was my hero and an inspiration for many things I have done with my life. He was the one who told me that I should make this field my choice as a career. He always had a way of showing you your true gift and potential. He always saw something in me that guided me to not only just do the job, but to do it well. I thank him every day for being the loving father that he was. I know with every patient I treat he is there with helping me through each call.

Walking around the compound I began to notice many things that really got me concerned, Bruce woke up this morning and told me he was done and couldn't do it anymore, telling me he was burnt out and misses his family too much. Suffice to say I was pissed off at him, but I can't make someone do something when they out right refuse to go any further. We have been here for 9 days and like him I was tired as well as were the rest of us and we all miss our families, but we all knew what we were getting ourselves into. We all knew we could be down here for week's even months and we accepted this, it's really kind of embarrassing that I could no longer rely on Bruce, my teammate, as he just wanted to lie on a cot and sleep. I was later approached by other medical personnel who were willing to work so as missions would filter in from EOC I knew that I was good to go.

Another ambulance service has come in to help (if that is what you want to call it) and has taken over operations and set up their own shop. They apparently got the FEMA contract for mass disaster response and this has left a lot of other rescue and ambulance services to pack up and head back home because the other service is taking all of the missions and leaving nothing for any of us except the crappy maintenance assignments that they don't want. I think it is really unfair for them to have taken over when they weren't here when the storm first hit. We are left sitting around doing nothing but sit and watch all of their ambulances drive in and out of the compound. They weren't here for the worst of it, the mud, the blood, and the overall misery of the whole situation, but they want to come in and take all of the glory while we stand here. I think about calling our bosses and telling them what is going on to see if we should just come back home now.

Talking to Kyle, we are going to stay for a few more days and wait for the other crews to come in before anything else happens. I told management several times through reports that sending other people may not be necessary but they want to send them anyway. It's frustrating to think that management is not listening to me or respecting my opinion or authority after all I thought that was one of the reasons why they sent me. I am right here in the middle of it while they make decisions from hundreds even thousands of miles away. It's funny how I can compare management to a woman, you don't understand them and the more you try the more confused you get. It's better to just know your limitations and just let it go.

Trying to see every outcome as a positive outcome no matter what the circumstances that brought you into it can be a difficult thing. We look through history and see that we can't determine and predict anymore, it's all-reactive and post plan to everything we endure. Being here has taught me a lot about my team, my field, and most of all myself. You look back at these events and you remember the good and bad times, the lives lost and saved, and most of all the people you stood side by side with to accomplish the mission. We achieved every objective that was set before us; we saved a lot of lives, and in some ways maybe even New Orleans.

I pretty much spent the cleaning and re-stocking the ambulance of used and depleted supplies, you just never get a clear picture of all the work you have done until you see how dirty and empty your ambulance is. I also spent the day cleaning and disinfecting all of our equipment and our clothes from the multiple chemical and other contaminants that we encountered. The continuous heat and high humidity has not helped with trying to get work done.

Tomorrow will be another day and hopefully we will be needed for something. It's when you go from full throttle to screeching halt you start to have the feelings of fatigue and frustration at the same time. It really messes with the mind and you just don't want to think about it. If everything continues on this course maybe we can all go home, I miss my family and the feeling of my warm bed. I know I we have ended all of this on a good note feeling proud of the team, the work we did and including myself. We came down here and accomplished our mission and people are alive to prove exactly that.

9/9 1120

AFTER A DAY of complete boredom and sitting around from the day we received pagers at 0830 only to have to return by 11am because they didn't work. I ended up taking a nap in the back of 9704 as I waited for the new guys to get here. Bruce, Jack, and Gary are heading back home this morning as soon as their relief comes in. Our bosses felt like the guys had worked hard enough and decided to have them come home, not to mention they weren't too happy with Bruce once I told them that he didn't want to do anything anymore. Last night Kyle and I decided to take Gary, Bruce, and Jack out to Hooters to say job well done and to just sit and talk about the things they had experienced and make sure they were good to go. The Hooters staff was great our waitress Ashleigh took good care of us not to mention having the staff take a picture with us was just another highlight. There was a table of folk next to us, who bought us drinks to say thank you for coming down.

Wherever you looked in the place we saw television screens with news about Katrina and it either made us want to laugh, cry, or throw something from sheer anger. There were the usual political leaders claiming racism speech about how the rescue workers were focusing on saving whit people rather than black people. Those sons of bitches had no right to make those statements, as they had no idea what we had been going through day in and day out. We (rescue workers) black and white, short and tall, fat and thin have always been focused on one goal from the very beginning, getting everyone out. I didn't see any of their ignorant assess in boats, in shelters, or in an ambulance losing sleep trying to all of the people out of that hellhole. They make their interviews and complain, but in the end they take no action to make a difference. It's amazes that even through all of this tragedy how people come from all over the world and try to make something positive only to have it destroyed by accusations of racism and conspiracy theories. It's based on a belief system that as a society we have come to live by, an individual can take something they believe in whether it be true or not and convince someone else that they are right you have started the system. These politicians took a belief of hate, racism, ignorance, and failure and planted right into the heart of New Orleans. The survivors and the sacrifices of the men and women in public services have been overshadowed by this negativity.

I look at the news and see and hear the words of body bags and body count and can't help but want to close my mind to the mere thoughts of it all. 25,000 body bags brought in for disposal and you think in reality of how many of those bags will actually have someone in it, and it makes you wonder if you really did anything at all. The families lost, the graves that will be filled and worse those that will never be found or even claimed. The places where the dead and living both suffered, the airport, the Causeway, and the neighborhoods, causes me think if we really did all we could. Do we look at it as a successful failure or as a failure made into success? At what point do we consider this a victory? We look at the folks that fought to escape and survive and we call that a victory, the family that is re-united and found safe, we call that a victory, the rescue worker that gives a stuffed animal to a child so they can smile that a victory. It's the small victories that make it worth while and build it into the big ones and soon after time when the sweat and tears have been wiped away the victory is complete with families back at home carrying on with their lives.

Each day will come a new challenge for us and we have grown as a group to be prepared for almost anything, a better New Orleans will come from this as we will all learn from these mistakes and hopefully do a better job in preventing something like this from happening again. Through events and testing yourself you either come to have faith or you lose it, maybe that's why the command post was where it was, a school of religion and faith. Reading the Bible and being around chaplains who have been through the same conflicts you have been through, they help you to understand of the how and whys of everything. You try to be true to yourself and the job you do and still face the hardships of death and suffering. Maybe that is what makes this job so fulfilling you face hell to make it to heaven.

9/10 2152

ITS GETTING FRUSTRATING to be just sitting around doing nothing, but it's even more frustrating sitting and waiting for someone to tell you where you will be going next. Kyle and I continue to go over the FEMA log sheets for daily events and it seems to be more and more blank with the passing days. Ryan and Frank have arrived and we have been going over the rules and protocols that have been given to everyone in the compound.

Ryan is a supervisor such as myself and will taking my place as one of the team leaders, I've known the man for 7 years now and he is a very trust worthy outspoken individual that can be a good thing to have on your side. A father of 3, a husband and an EMT he is driven and motivated like most of us and just wants to see the job done right.

Frank is pretty much a taller, older and thinner version of Ryan and is just as outspoken as Ryan is, he can be a cranky guy who I think swears and yells in his sleep. Frank a single father and paramedic and he will be taking Kyle's place as co-team leader and is looking forward to getting his hands dirty.

I try to keep myself busy walking around the compound talking to those who I have worked with to see if they have been having the same delays as us and it is pretty much the same story every where I go. They have all worked and dedicated so much of themselves and worked harder than anyone I know and have the same feelings as most of us. They just want to finish what they started and just wanted to hear the words telling them they have done their duty and have earned the right to go home. The waiting game has not been our friend it allows other emotions to clout judgment and lose motivation.

Hundreds of people including myself have had our minds set with endurance and guidance to saving New Orleans and all aspects of the primary objectives. Those who wanted to leave have left and now it's just a matter of keeping those who stayed behind safe and alive. Looking around at those who are here and you see the same face knowing the mission isn't over.

There are parts of New Orleans that will forever be changed and for some the recovery transition my never happen. The worst hit area the 9th ward has been completely wiped out with houses washed away or just uninhabitable, bodies floating in the water, and cars submerged in the water like submarines. They finally started to retrieve the bodies of the elderly who were left to die in their bed at several nursing homes. The staff evacuated and left the patients there either under orders of just of their own lack of compassion and common sense. There are very few things that compare to leaving someone to die a very slow and cruel death while you selfishly think of yourself and flee. We were under strict orders for a few days not to remove those bodies until specialized team came in to investigate the nursing homes because they were considered crime scenes. Someone needs to be held accountable for the deaths of these people, those who couldn't help themselves and relied on their ability to care for others. It will only be a matter of time before everything comes to pass and it will.

I think of the friends I have made here and know of the sacrifice that they have all made to be here, so many different people from ambulance service all over the country, hospital nurses and doctors, and regular staff who have sacrificed just as we have and doing it without a second thought, it's a true bond of dedication and love for the job that you cannot break. I feel myself coming close sometimes to the breaking point, like I can't take much more of the madness, I feel so drained mentally and physically all of the time and no matter how much sleep I get I always feel tired. I have been here so long it feels like its been months and it's been just a couple of weeks. I sleep and I can hear the screams of those who are in pain and dying and I wake up in a cold sweat with tears in my eyes. The men and women, young and old, and most of all the innocent children who didn't ask for this misery and nightmare, as a whole we owe it not only to them, but to ourselves to save their lives and the lives of their loved ones. I think about Munique and my kids, how much they mean to me, how much I love, miss them and can't wait to get back to our lives as a family.

The compound has also gone through some changes with security the National Guard has brought in a couple of soldiers to patrol the grounds to ensure that nothing happens. After all of this time, the confusion and people running around everywhere things are calming down they decide to bring in these guys. The only problem is none of us have any type of FEMA identification to say that we belong here so they have no way of knowing who anybody is. EOC and FEMA seem to have been one step behind this whole time with everything from the very beginning until now and it has become quite comical. It doesn't seem to make much sense, but being down here and seeing what we have seen, nothing seems to make any sense anymore. It's like the concept of wiping your butt before you go to the bathroom, there is nothing accomplished in doing this.

9/11 1049

FOUR YEARS AGO today America endured one of it's darkest days of tragic loss and destruction and today here we stand enduring another day of tragic loss and destruction. Although we are not the same brave and gallant men and women who worked and sacrificed, we only hope we could bring the same closure and strength to the world as they did. Their strength, determination, and courage, was the glue that bonded and unified a country to conquer any odds.

This morning I woke up thinking that I would be evacuating a hospital, but instead I'm at the floodwaters waiting for rescue crews to bring back any potential survivors. Before we left Kyle and I were going to split up Ryan and Frank so we could have them exposed to streets and operations, but Frank threw a big tantrum about everything so Kyle and I took the assignment. After sitting around for three days and not doing anything, it felt good to be around other people who were trying to do the same as us.

This part of the operation was being run by Los Angeles Fire Department is one tight, disciplined, well oiled machine. Being around them even reminds me of being back in my hometown of San Diego and talking to them, they even grew up around the same parts as I did. These guys have a complete system of command and other joint operations working together and it is great to see. It's nice to see some true communication and organization with no attitude or egos pushing themselves around. It's pretty much a standing rule of "You do it our way when we tell you or don't come back."

The water is so nasty it's been giving off this foul smell, not like the smell of death, but something else it's like being around a garbage dump, the water is black one minute and then like a hunter green the next. The National Guard has been doing tests on the water trying to find out exactly what is in it and there are chemical compositions that cannot be identified. The thing that has been the most terrifying is the people have been walking around in this stuff surrounded by everything and anything. There have been rumored reports of the search and rescue dogs dying from going into the water, now you can just imagine how toxic that water is to people also. Being here I think that we may find something or we may not but it's important for us to know that we are here doing the best we can.

Looking around we were able to see how high the water actually was when the area first flooded, the water level looks like it has dropped 6-8inches since we have been here and continues to slowly drop. Many of the one-story houses still have water up to their rooflines, but some of the two story dwellings may have survived. Many of the cars are still submerged as others look like fishing bobbers on the water's surface. The weather is still hot and humid as the rescue crews continue to patrol around houses making sure no one was inside, there must have been about a dozen boats and jet skis just doing what they could.

Kyle and I decided to post in the back of the ambulance and just sat back and watch what's been going on around us. This triage/medical center is ready for anything as they have decontamination showers, medical team, a station for animal rescue that had several dogs and cats in kennels ready to go to a better, safer place. Being around all of this has shown me what dedication and commitment really is about, the job we do, saving everybody and everything, whether it be animal or human if we are able to rescue it, we have done exactly that. It's through the training, teaching, learning and applying that the following applies, "It is better to have and not need, than to need and not have." In his case we got everything just to fulfill the need to better care for those who cannot care for themselves.

9/12 0757

THERE HASN'T BEEN a whole lot going on in the last few days as the missions have become fewer and farther in between. It seems to be luck of the draw of what personnel get to go where and for how long. Some of the missions are requiring personnel to be isolated in one spot for 24-48 hours and that can cause a lot of physical and mental strain on these crews. Still through it all we have accomplished more in the past 13 days than most people have done in their entire lives. We have saved a lot of lives and prevented further harm and danger to the people of New Orleans, in a way we could call ourselves heroes as a lot of people have, but to be honest I don't feel like one. Thinking about it, I couldn't honestly tell you what I think a hero is. We see a perspective of them in movies and even then you wonder what actions define a person to be a hero?

I have looked around this compound and there is stillness now, the evacuation is 98% complete and the floodwaters have started and continue to recede. You look back and think about what happened here and you think about the people that will remember all of this. It is the type of event that requires reflection on what went wrong? Unfortunately it will be as a society that is how people remember what to do correctly the next time, but hopefully there won't be a next time. Is it the type of event that requires us to reflect on what we did right? Absolutely, it helps to build a foundation of retelling the story the way it should be told. We walked into a situation that seemed to have no positive outlook of hope or success and we came out with an honest chance of being able to start over. It's seems through history a tragedy that seems hopeless can rebuild itself through the spirit of people coming together and doing wonderful things. It's baffling how you can take a group of people from different walks of life from all over the world, put everything aside and pull off amazing things.

9/12 1055

I LOOK BACK at the time I have spent here and all I do is think about the hell that others and myself have endured. I think of the 7 month-old little girl that we rescued from the Causeway, we kept her safe and kept her alive so she can grow up with her family and do something important with her life. I have these images and many others that will be the demon s my burden and the same burden of many others here endured. The faces of desperation and devastation on the people who have nothing and in some way I can understand and relate. Seeing the dead floating lifeless in the water knowing that they were part of someone's life, the pets seeing them scurry around trying to find their homes and families, houses flattened or severely damaged, windows smashed and blown out, traffic lights toppled, and the water that ravaged and destroyed lifestyles and life itself.

Mother Nature's rage had no boundaries, no mercy, and through Darwin's own theory of natural selection was made. It makes you wonder and question God and why He has done things He has done, question Him and He doesn't give answers, but when you are full of rage, anger, and hate you really don't want answers. For some justification you just walk away from God and refuse to believe in any faith and become almost empty inside.

Reading the Bible I can't help but go to the Revelations where the end of the world seems to be pertinent to now. Revelations 6:8 speaks of, "And I looked behold the pale horse and a name that sat on him was Death and Hell followed with him." The irony of it all is so thick as we the earth were the pale horse, Katrina was death, and hell was floods and ruins that were left behind. It feels as if my belief in the love of the Lord being betrayed by Judas in John 12:4. Was what we had done as society and earth so sinful that the Lord would promote so much death?" I look for answers through His word and all I find more reasons to question my faith in Him. He is my Savior as I learned as a child, but how vengeful must you be to inflict so much misery to one's soul? It's not about revenge, it's about knowing and believing what is right and wrong.

Slowly day by day more and more personnel are leaving to head back home from a job well done. In a way I envy them as I look forward to when I can get home, the idea of being able to hold my family is something I think about everyday. I think about those who are separated from their loved ones because of all of this, those missing, hurt or passed away all the things they will miss like weddings, anniversaries, graduations, and all the other occasions for togetherness and celebration. The homes that were ravaged and burned, the buildings that collapsed and the lives permanently changed from all of this.

Even after the clean up and rebuilding nothing will ever be the same. No matter where you look, no matter where you stand it will always be there. The unforgettable images of how New Orleans, Biloxi, and the other areas were devastated Katrina.

The sights, the smells and even the feelings will always be part of us, like a picture etched in our minds never to be forgotten. Maybe that's a good thing, way to have us remember how one event can affect so much of the world around us.

9/13 0830

AFTER 12 LONG exhausting days and longer sleepless nights we (Kyle and I) are on our way home. It's been a long hard journey we have taken and I know this has forever changed me. It will take a long time before I will be back to a routine of daily life. It was hard to leave last night, the feeling of family and unity in New Orleans was very strong. Down here all business rivalry, personal animosity, and bitterness was gone to provide for a common goal, survival and rescue. We had one single heartbeat, a single thought that provided enough force to make a formidable dent and impact on the ravaging that Katrina had done. There were tears and goodbyes as we left the command post that day, I feel closer to these folks in 12 days I've spent in New Orleans than all my years in EMS.

We got so caught up in talking that I didn't even notice how fast I was driving and ended passing a state trooper going probably close to 80 mph. I knew he was going to pull me over so I slowed down and pulled to the shoulder and waited for him to pull up behind me. He asked if I knew why he was stopping and I told him because I was speeding and the obvious look of "no kidding" crossed his face. He asked me where I was coming from and when I told him where we had been, he told to slow down and sent me on my way. Kyle and I just started laughing hysterically at the whole thing and how stupid I was for pulling over and waiting over for him.

Part of me wanted to stay because I felt like I hadn't completed the job, but I knew I had done so much already and I served my tour of duty and walked away with a true sense of accomplishment. I will never forget those I stood side by side with in good times and in bad. We cried on each other's shoulders, laughed, vented our feelings, and worked feverishly to help the folks who lost everything. I think that no matter what was going on whether it was good or bad all of the people down there had a code of honor and brotherhood, "we go in together and we walk out together, no matter what everyone goes home alive." I think about the world we live in and all the hypo racy, it's the "I have your back when I have something to gain" or it's the ever classic, "I'll say I have your back but I'll ditch you when it comes to the real test." If we had been like that in New Orleans none of us would have came home with a sense of pride, we would have screwed up and someone would have ended up dead or seriously injured.

I think about those I have shared this event with from both near and afar, those I will always call my friends a new family that was created. We were in a war, but without guns or tanks, it was a war of time and often times it did kick our ass there is no denying that, but we always seemed to come out on top. All of them Tim, Jim, Eric, Mariah, Becky, Britney, Brad, Tom, Jaime, Julie, Kim, Ryan, Matt, Adam, Sharon, Steve, Debbie, and just so many other people who came in from near and far, so many stated and countries who answered the call. We all put our lives aside for others and looking at it from the job angle, it's what we do. This is what the job asks of us, we save lives.

On the way back, our bosses were telling us how they have news media and a lady from a local paper coming in to interview and talk with us about this experience and I'm not sure if I'm wanting to do that. My opinion of media has never been very nice ever since back from 9/11 to especially now it is not held in high regard. The other thing is, I don't want to us to be the spokesman for what went on in New Orleans, it was about the people who lived there, not a bunch of guys form Pennsylvania who wanted to go and make a difference. I may not even be ready to talk about this to people I know our story needs to be told and people need to know the truth, but I am not sure if we are the ones who should tell it. I wont speak for them; they have their own voices and have the right to say something or nothing at all. Right now all I can think about it hugging my family, having a home cooked meal, and being able to sleep in my own bed.

There are a lot of small things that trigger memories in my head about New Orleans, whether it be seeing a boy waving at me as his family's car passes by or even hearing a song and listening to the words. There was a song I was listening to and it made me think of all of the people we were around. The words sang of someone looking at someone and not being able to take their eyes of someone while other people surrounding him with nothing to do, but just stand there. It felt like the song was hitting me right in the face as if they were singing about all of us. It makes me think of the evacuees/refugees and what they are going through as we were just trying to help. They can't take their eyes off of the reality of this tragedy and they don't know what to do. The symbolism through all of this is we all have nothing to prove to anybody, but ourselves and being there we came in and did the job and proved how strong we really are. We took a situation that seemed dark and grim and turned it into a unified fight to victory that no one can ever take from us. Mother Nature had our butts kicked, but in the end the battle was won by us, we never gave up.

While we were there, I got to thinking about the events of 9/11/2001 and it seemed morbidly ironic that we were pretty much doing the same thing 4 years later. Men and women dug through concrete, steel, and dust to find people who were the innocent victims of a mad man's rage. 4 years later here we were digging through mud, water, wood and concrete to find the innocent victims of Mother Natures rage. It's a strong presence of mind to see members of the New York Fire Department working side by side with us. To be able to put aside the grief they still carry from the ones they lost and put on the uniform, hop into a truck and come to New Orleans is an unwavering sign of courage, determination, strength, compassion, and brotherhood.

9/13 2342

AFTER A LONG drive home we reached the base to find a lot of people waiting for us with two news crews ready to give us interviews. I was wondering why our bosses kept calling Kyle wanting to know where we were and who was driving. Our bosses wanted us back pretty quick and with my lead foot getting there wasn't a problem. Immediately after we exited the vehicle, Zachary my one son ran over to me and gave me a big hug. The news camera was in my face but I just shrugged it off so I could hold him. My other son Dylan was running around somewhere and didn't even know I was back. I was looking forward to hugging my fiancé Munique, but she was sent on a call a few minutes before I got back.

The news lady came up and stuck a microphone in my face and wanted an interview and I hadn't even caught my breath yet. With all of my bosses in front of me I didn't want to cause an issue so I went along with the show even though hatred towards the media was well known by my bosses. The first question she asked was what was it like there? I didn't really know how to answer the question at first, I mean how do you look at into this person's eyes and into a news camera and tell them the hell you were just in? The Superdome, raping and beating of women and children, looting going on by civilians and public servants caught on camera, police officers and other public servants abandoning their posts or committing suicide when they hear that their families are dead, dead bodies found in the street and not being able to do a damn thing about it, finding thousands of people at a shelter needing help as they die by the hour, parents separated from their children, families suffering the death of a loved one, and most of all seeing the press and public figures point figures of blame when they should have just shut up and came up with a solution. You just smile and try to tell the story without losing yourself in the madness and still be able to go home to your family. When I was done with the interview, I hugged my team, grabbed my gear and got out of there as quick as I could. I really didn't expect all of this nor did I really want it at the same time, I was just tired and wanted to go home to get some sleep. I'm just glad to be home.

9/15 1056

BEING BACK HOME and already back to work I was reading over my notes, I broke down and it really pissed me off, telling the stories of the mother trying to care for her children by beating of angry people so she can have a cot to sleep on with her kids or the man who tried to kill himself by lying under a bus because his family was dead.

Looking back at it all of this I realize how much I took for granted and how much we all took for granted. We have a simple job to do, save lives and go home. Screw the politics and details that try to control how we do the one thing we are trained for. I came home to move forward, tell our story, and have the world know what we did and saw. People died, we couldn't control that, but we were able to take what we had, made the best of it and saved a lot of lives. There were nights we slept on a cot, ate some nasty food, didn't shower or brush our teeth and in some cases wore the same clothes for days. We however were alive when so many weren't; we had families and homes to come back to when the job was done there were thousands who didn't have anything. I am blessed truly blessed to be as fortunate as I am. For those who complain about the stupidest crap, I say go to New Orleans and endure what they did and then complain.

9/15 1508

I STILL FEEL angry with myself, God, my bosses and most of the media and how they have continued to handle New Orleans and the struggles they face. I want the truth to be told in black and white realism, as it should be, there were no heroes, and just those who chose to do something about it, the ones who gave up everyday life to be there for another human being. We were there when they called upon us and we gave everything we had to help each and every person we came across. Now as we came home we can still hear the screams in the night of those pleading for help.

There were some we had to leave behind because there was nothing we could do for them; there were those who refused to leave, as there were those who couldn't leave for whatever their reasons may have been. Some died without a chance as the floodwaters came at them too fast for them to escape, as others died just sitting and waiting for help. Some died with no hope of rescue in sight no boats, no helicopters, and no signs of life. You can never feel as much of a failure as seeing death right in front of you and feel completely helpless

The tears never stop flowing, like the waters that flood they just recede to another place in time, the nightmares never go away, they just rise and fall as the sun. Forever the images encased in our souls as if these images were a photograph that never fades.

9/17 1518

WATCHING TELEVISION TODAY I was able to watch CNN and see that there were still heavily flooded areas and a chill went up my spine. They were showing pictures of kids that were listed as missing and three possibilities of their outcomes came to me: 1 they were evacuated/rescued and shipped off separated from their families, 2 their family doesn't know where they were when the storm hit, 3 they are dead and the families haven't been notified or the body identified.

There are so many missing or dead some of them will never be found, having have been eaten by animals, the water and all the chemicals have destroyed the body, or have been carried back to the open water by the current. Reports have the body count at 700, but they haven't started out a full-scale body recovery yet either. There are parishes (towns) that have not been searched for survivors or bodies because once the levees broke and there was no way anyone could get to them.

I still find myself waking up from the images I have, in a cold sweat, shaking and trying to catch my breath, but then I don't remember what it was I was dreaming about. I have woken up from a sound sleep in the middle of the night by my own screams, my arms and legs weak and tired from my body flailing everywhere. Things like that make me concerned for my safety and the safety of my family. Will this ever end? How long will this continue? I pray to God for the closure I need so I can finally get a peaceful night sleep.

I've been through the stress debriefing session, but it didn't do anything, the guy couldn't have given a rat's ass less about how we felt or what we had to say. I honestly don't know what to think or feel, I am physically and mentally numb. I almost feel empty and hollow wanting something there inside me to make me feel like a person.

9/18 1000

FINALLY THE FIGHT was coming to light as the view of racism was once again brought to the world's attention; only this time people have verbally voiced their displeasure at these accusations. Several public figures tried to tell the world how African American people were segregated in rescues. Many people have gone public and blasted these people for their comments of ignorance and lack of respect, support and overall humanity. They publicly criticized our efforts to rescue those lost, when I didn't see these same folks in a boat to lend a hand or working for 20 hours straight in a triage center. We were seeing no recognition from the media or public figures/officials to the men and women who traveled near and far to help.

We didn't play the racism game; we were playing the beating the clock game for people's survival. There were people who refused help and there was nothing we could do, when someone died no matter who they were, it affected us. In our eyes, death was not an ethnic selection, it doesn't matter what color you are, how much you have, who you are, or what name you call God death took you and every one of them affected us.

Seeing a young kid in the water just floating away it tore me apart knowing that I couldn't do anything. A young life was taken from this world and his skin color didn't matter, it sucks the sheer ignorance and stupidity that has come out of people's mouths in front of a camera. It's not about the truth or a lie it's about what they could get people as a general public to believe.

I remember being at one of the rescue sites with boat crews hearing of a man looking for his family, the rescue crews found his family in the house dead, it didn't matter who this guy was or who his family was, they were dead. There were no favorites played, no one was picked over another on basis of any race, color, or creed we simply picked life over death.

9/19 2356

TONIGHT I WAS watching a movie that I have always enjoyed, about the brotherhood of soldiers during the Vietnam War. A journalist said about his own experience in Vietnam, "We who have seen war never stop seeing it, in the silence of the night we will always hear the screams so this is our story." I know that Katrina was nowhere near in comparison to the Vietnam War and in no way try to compare the two, that was a true war and the men and women who fought truly are heroes. I do understand what the man says it that statement. I have been to places where there were screams, screams of pain, agony, sorrow, and impending death and you that sound never leaves you. It's the never ending nails down the chalk board that sends a sickening feeling into you and a shiver down your spine.

For New Orleans Hurricane Katrina left the screams in the stillness of the night, the screams of the sick and dying right in front of us. The children crying from hunger, the begging for help by a single mother for someone to save her children, the loud prayers of a father wanting his family to be rescued, the cries of pain from those hurt, and the silence of those who died 10 feet from us and we couldn't help them. Having to look in the eyes of a person and watch the life literally drain from them is something that never leaves you.

Those who yell from heaven, their souls unrest from their tragic death, their empty shells that float on the water's surface, the type of scream that is so pure and distant that it kicks us in the soul of our being. Having to leave them behind to find those who still have breath, they may not scream or cry to heaven, but it's still a sound that wakes me in the middle of the night and may do so for the rest of my life. For man of us this may never end yet still we all accepted our fate without regret.

To this day the events of Hurricane Katrina still affect us all as to the mistakes that caused so much damage and cost so many lives. Despite all that went wrong it is important what was done to turn it around and make it into what we believe as a successful mission. The men and women of EMS, Fire, Law Enforcement, and the other resources who came when New Orleans called this is a tribute and dedication to the work we do.